JOY OFFICE DESIGN II
怡悦·办公 II

PREFACE 前言

办公空间设计是一门实用艺术。随着都市工作节奏的加快、办公时间的加长,设计一个舒适、轻松、愉快而生活化、人性化和智能化的办公空间已成为新办公空间的设计理念。目前国际上较为流行的办公空间设计形态大致有四种类型:蜂巢型、密室型、小组型及俱乐部型。蜂巢型办公空间是典型的开放式个人办公空间,传统工作流程、蜂巢式配置,自律度及互动度最小,适合朝九晚五或者24小时轮班制的工作模式;密室型办公空间的表现形式为个人化的独立工作空间,其自律度高、互动度低,工作时间及地点较不规律,适合研究性工作;小组型办公空间适用于在喧闹环境中的小组办公和交互式团队办公,工作形态属于工作效率较高的团队小组式;俱乐部型办公空间属于个人和团队合作的模式且经常需要小组讨论,适合于高自律性和高互动性的知识工作者。

不同的企业由于业务类型的区别及办公室的资讯化发展,对于办公空间设计的需求也不同。这就要求设计师对周边环境、空间布局、色彩搭配、光影应用、家居摆设、软装配饰等各个环节进行深入的思考,得出合理的设计方案,一方面营造艺术的美感,另一方面创造完整的功能分类,给使用者带来便捷、舒适的使用感受。好的设计是艺术美与实用性的高度结合,因此对设计师们提出了更高的要求,不仅要有良好的美学修养、积极的创新能力,还应兼备国际化视野。

本书精选出51个代表全球最新潮流的精彩办公空间设计案例呈现给大家,囊括了欧美、亚、非、澳等全球各个地域最新不同风格的作品,充分代表了当今室内办公设计领域的最高成就与发展趋势。从天马行空的设计公司到思维严谨的法律事务所,从世界500强的大型公司到半SOHO方式的个人工作室,涵盖各个领域的各种风格,全面而精练地向您展示了针对不同企业和不同面积办公空间的解决方案。每一个案例各具特色,均配有完整高清彩图及平面图,不仅可以为高端室内设计人员提供专业参考,还可以为一般室内设计相关人员和高校专业师生提供借鉴。

The office space design is a practical art. With the acceleration of the city's working pace and the extension of office time, designing a comfortable, relaxed, happy, life-oriented, humanized and intelligentized office space has become the design concept of new office space. The current international popular office space design form mainly has four types: honeycomb type, back room type, group type and club type. The honeycomb type office space is a typical open personal office space, featuring traditional workflow, honeycomb type configuration, low self-discipline and little interaction, which is suitable for nine-to-five or 24 hours shift work; the back room office space is an individualized independent working space, featuring high self-discipline, low interaction, irregular working time and place, which is suitable for research work; the group office space is suitable for group work and interactive team work in a noisy environment, whose work form belongs to high-efficiency team work; the club office space belongs to the type with individual and team cooperation combined, which often needs group discussion and is suitable for high self-disciplined and high interactive intellectual workers.

According to the different business fields and the informational development of the office, different companies have different requirements for the office space design. It requires the designers to take surrounding environment, space layout, color matching, light and shadow application, home furnishings and decorations, soft decoration adornment and so on into deep consideration to get the reasonable design plan, which on one hand creates a sense of artistic beauty and on the other hand creates complete functions to bring convenience and comfort for the users. Good design is the high combination of artistic beauty and practicality. Therefore it puts forward higher requirements for the designers who not only should have good aesthetic cultivation, active innovation but also have international vision.

The book selects 51 wonderful office space design cases representing the latest design trend for readers, including the latest different-style works in each area all over the world like Europe and the United States, Asia, Africa, Australia, which fully represent the highest achievement and development trend of the interior office design today. From the powerful and unconstrained-style design company to rigorous-thinking law firm, from the world's top 500 large companies to half SOHO personal studio, this book covers various styles in every field, which shows the office space case solutions for different companies with different areas comprehensively and refinedly. Each case is unique with a full HD pictures and plans, which can provide professional references, not only for high-end interior designers but also for general indoor designers and college professional teachers and students.

Contents 目录

- PORT OF PORTLAND HEADQUARTERS AND LONG-TERM PARKING GARAGE — 008
  波特兰航空港总部与长期停车场
- TWELVE WEST — 016
  十二西大厦
- OFFICE OF BANK OF MOSCOW — 024
  莫斯科银行办公室
- ATTRACTION MEDIA OFFICE — 032
  Attraction 媒体集团办公室
- PARSONS BRINCKERHOFF — 040
  柏诚集团
- ADIDAS WORKOUT — 046
  Adidas 德国总部
- SINCLAIR KNIGHT MERZ PTY LTD. — 054
  Sinclair Knight Merz 公司
- PAGA TODO OFFICES — 060
  Paga Todo 公司办公室
- YANDEX KIEV OFFICE — 066
  Yandex 网络公司基辅办公室
- KNOCK INC. — 074
  KNOCK Inc 公司
- 1-10DESIGN KYOTO OFFICE — 080
  1-10design 网页设计公司京都办公室
- GUOGUANG YIYE XIAMEN BRANCH — 086
  国广一叶厦门设计分公司
- YANDEX MOSCOW OFFICE — 094
  Yandex 网络公司莫斯科办公室
- FORWARD MEDIA GROUP PUBLISHING HOUSE OFFICE — 100
  前锋传媒集团出版社办公室
- ZHENCHUAN SHOP BY XUPIN DESIGN LTD — 106
  叙品设计震川店
- DRAGON · TRIPOD — 112
  龙·鼎
- ELEMENT · PLAIN COLOR SPACE — 118
  素色空间
- MCCANN-ERICKSON RIGA AND INSPIRED OFFICE — 122
  麦肯埃里克森里加创意办公室
- MCCANN-ERICKSON RIGA PR-AGENCY — 128
  麦肯埃里克森里加公关部办公室
- FRAUNHOFER PORTUGAL — 134
  弗劳恩霍夫办公室
- VIP WING IN MUNICH AIRPORT — 142
  慕尼黑机场贵宾休息室
- SIGNA HOLDING HEADQUARTERS — 148
  思歌娜控股集团总部
- KUSTERMANN PARK — 152
  Kustermann 园区
- LHI HEADQUARTERS — 158
  LHI 总部
- NEW OFFICE DESIGN ICADE PREMIER HOUSE — 166
  ICADE 总理府新办公楼
- INTENTION · PRIMARY COLORS — 174
  初·原色

- MPD OFFICE
  MPD 办公室 ... 182
- CHONGQING CHUANGHUI FIRST OFFICE MODEL HOUSE
  重庆创汇首座大厦办公样板房 ... 186
- YBRANT DIGITAL
  Ybrant Digital 公司 ... 190
- NEWPORT CITY CAMPUS, UNIVERSITY OF WALES
  威尔士大学新港城校区 ... 194
- ZHONGQI GREEN HEADQUARTERS · GUANGFO BASE OFFICE
  中企绿色总部·广佛基地办公室 ... 198
- PARFUMS & BEAUTÉ
  PARFUMS & BEAUTÉ 公司 ... 204
- ASTRA ZENECA HEADQUARTERS
  ASTRA ZENECA 总部大厦 ... 210
- UAWITHYA CORPORATE HEADQUARTERS
  Uawithya 公司总部大楼 ... 214
- ZAIN HEADQUARTERS
  Zain 总部大楼 ... 220
- A.M.A. HEADQUARTERS
  A.M.A. 总部办公楼 ... 224
- SUGAMO SHINKIN BANK / SHIMURA BRANCH
  日本东京巢鸭信用银行志村分行 ... 228
- ARCHITECTURE OFFICE—LOFT
  LOFT 建筑办公室 ... 234
- CORPORATE OFFICE FOR APOLLO TYRES
  阿波罗轮胎公司办公室 ... 238
- BENE FLAGSHIPSTORE
  Bene 旗舰店 ... 244
- CREDIMAX HEAD OFFICE
  CrediMax 办公总部 ... 250
- BOUYGUES TELCOM TOWER
  布依格电信办公总部 ... 254
- FLEXIBLE SOLUTIONS FOR DESIGN OFFICES
  Design Offices 的活性办公空间设计 ... 262
- NEW HEADQUARTERS FOR HARLEY DAVIDSON
  Harley Davidson 新总部大楼 ... 266
- ESSENT HEAD OFFICE
  ESSENT 办公大厦 ... 272
- DUNMAI OFFICE
  Dunmai 办公室 ... 278
- BPGM LAW OFFICE
  BPGM 律师事务所 ... 284
- AND-SUPERPRESS-SUPERBLA
  AND—SuperPress—SuperBla 办公室 ... 290
- KOZA HOLDING HEADQUARTERS
  Koza 控股总部 ... 298
- ROTHOBLAAS LIMITED COMPANY
  RothoBlaas 公司总部办公楼 ... 306
- RED BULL AMSTERDAMR HEADQUARTERS
  红牛阿姆斯特丹总部大厦 ... 310

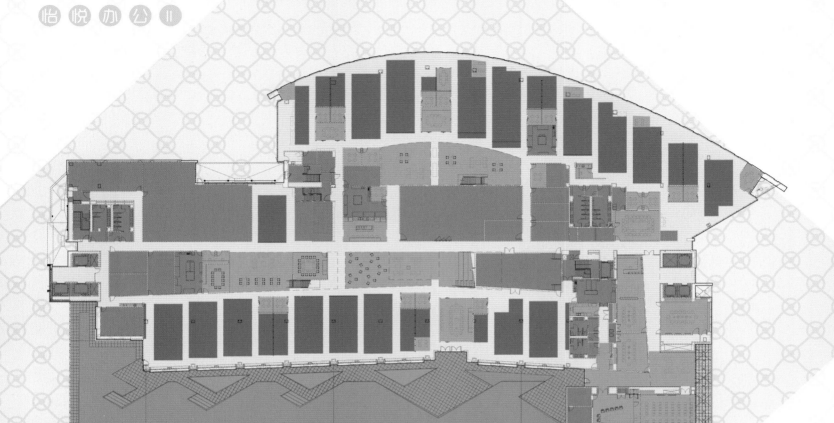

PORT OF PORTLAND HEADQUARTERS AND LONG-TERM PARKING GARAGE

波特兰航空港总部与长期停车场

ZGF Architects LLP

建筑机构：ZGF建筑事务所
建筑面积：19 100 平方米
项目地点：美国俄勒冈州波特兰市

Architects: ZGF Architects LLP
Area: 19,100 m²
Location: Portland, Oregon, USA

The 10-story headquarters building is situated to the east of Portland International Airport's main terminal building and connects to the existing parking structure, serving as a new gateway to the airport. The building is visible to passagers arriving at and departing from the airport, either by roadway or by air.

The Port of Portland's commitment to sustainable practices is showcased in its new headquarters building and long-term parking garage. Sustainable design is a critical driver to gain LEED Gold Certification. In addition, the building allows the Port to consolidate most of its workforce, bringing together approximately 240 employees. This will save additional operating costs and provide the Port with an opportunity to lease the airport terminal space to other users. It also adds parking capacity to Portland International Airport.

The new 205,000 m² building consists of three floors of office space atop seven floors of public, airport parking. The new offices reflect a 21st century's culture—"One Port"—in an effort to increase collaboration and foster a team environment. Cost-effective solutions were a primary concern, as well as telling the Port of Portland story through environmental graphics and artworks. ZGF worked closely with the Port to develop new standards for office space to accommodate a shift from a closed office environment to a primarily open plan.

高达10层的总部大楼位于波特兰国际机场主航站楼的东侧，并与现有的停车场相连，形成了通往机场的新途径。无论是乘车还是乘坐飞机，乘客都可以在到达或者离开时观摩这座标志性的建筑。

波特兰航空港总部大楼和长期停车场的项目不断实践可持续性设计，这是可以荣获 LEED 黄金认证的决定性因素。此外，这一大楼利于航站楼大多数工作人员的整合，将这里的 240 名员工汇集到一起。这将节省额外的运营成本，此外还将空置的航站楼空间租给其他用户，并扩大了波特兰国际机场停车场的容量。

面积为 20.5 万平方米的新大楼包括机场停车场、7 层的公共空间及位于其上的 3 层办公空间。新总部大楼反映了 21 世纪的文化精神，即努力完善协作、推广团队环境，打造"一体化机场"。设计主要关注如何有效节约成本、如何使用环境标志设计和艺术品来展示波特兰航空港。ZGF 建筑事务所与机场紧密合作，制定出办公空间设计的新标准，以适应办公环境从封闭到开放的转变。

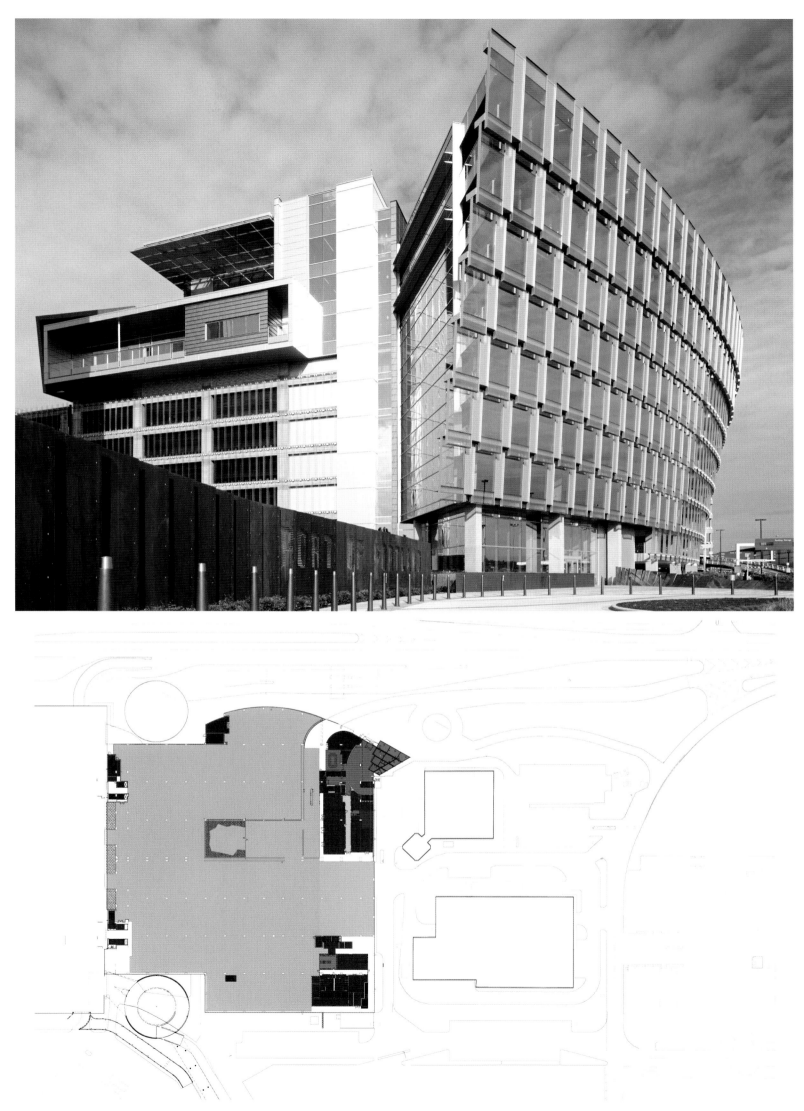

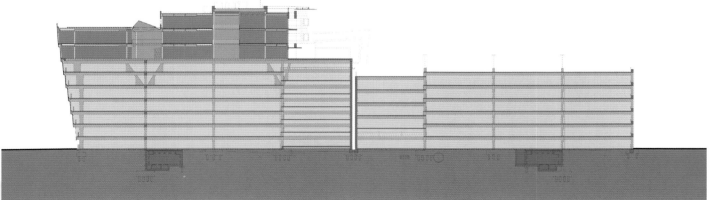

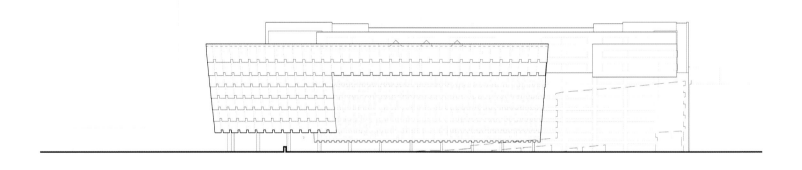

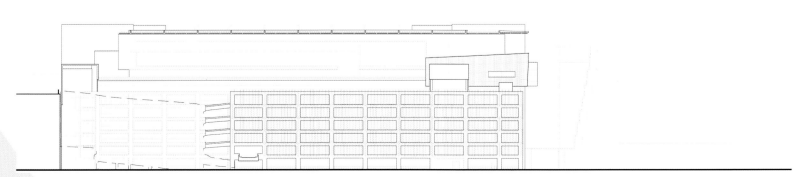

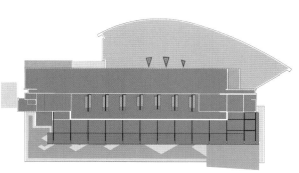

ground floor

- lobby
- retail space
- storage
- apartment lobby
- elevator/stairs
- neighboring buildings

floor 3

- office space
- public space
- conference room
- bathroom/locker room
- elevator/stairs
- outdoor space
- library

TWELVE WEST
十二西大厦

ZGF Architects LLP

建筑机构：ZGF 建筑事务所
建筑面积：51 100 平方米
项目地点：美国俄勒冈州波特兰市

Architects: ZGF Architects LLP
Area: 51,100 m²
Location: Portland, Oregon, USA

The concrete structure of the building is exposed on the interior whenever possible to provide thermal mass that will help provide a flywheel effect for indoor temperature fluctuation. The choice of concrete also means that the most ubiquitous materials in the building come from highly-recycled and local sources.

Additional recycled or local materials include: salvaged wood; high-recycled content and urea-formaldehyde free MDF cores for solid core doors, casework, and painted trim; recycled blue insulations and 96% recycled, locally-manufactured gypsum wallboard throughout the building. Rapidly renewable or sustainably-harvested materials include: bamboo veneers for doors and casework, bamboo flooring, FSC-certified wood, and 100% corn fiber curtains and linoleum flooring.

Twelve West is a mixed-use building designed to receive two LEED Platinum Certifications and serve as a laboratory for cutting-edge sustainable design strategies. It features street level retail space, four floors of office space for ZGF Architects LLP, 17 floors of "eco-chic" homes for lease and five levels of parking. The building has an eco-roof, rooftop garden and terrace space, complete fitness studio and a theater. Four visible wind turbines sit high atop the building representing the first installation of a wind turbine array on a high-rise in America.

Twelve West serves as not only an anchor in a rapidly transforming urban neighborhood, but also as a demonstration project to inform future sustainable building design. Twelve West is on track to achieve two LEED Platinum certifications—one under New Construction and a second under Commercial Interiors for ZGF's offices.

Because the lower floors of the building are dedicated to commercial activities, all of the building's apartments rise above the surrounding rooflines, offering unobstructed views in all directions. The 273 Indigo in Twelve West apartments, which begin on the 6th floor, offer a mix of studios, one-bedroom, two-bedroom and three-bedroom units ranging in size from 50~197 m², and include penthouse homes of three floors.

Modeling predicts that energy efficiency strategies utilized in the building will reduce consumption of energy by more than 44% and exceed the 2030 challenge benchmarks for this project type. Some of the efficiency measures include: thermal mass; daylight and occupancy sensors; low-flow fixtures for reduced domestic hot water demand; high-efficiency equipment; heat recovery; fan-assisted night flush of the office floors; chilled beams and under floor air distribution in the office floors; CO_2 sensors for ventilation demand control in large volume spaces.

此建筑内部显露在外的混凝土结构可以持续提供蓄热体,这将有助于产生调节室内温度波动的飞轮效应。建筑中使用的混凝土是最普遍的建筑材料,均来自高效回收利用的当地材料。

其他回收利用的当地材料包括:废弃木材;用于制作实心门、台柜和油画装饰框的、不含尿素甲醛的中密度纤维板核心部分;可循环利用的蓝色绝缘材料以及遍布整个建筑、当地生产的石膏墙板(其中96%采用回收材料)。快速可再生能源或者可持续利用的材料包括:用于门和台柜的竹胶合板、竹地板、由联邦科学委员会认证的木材、100%玉米纤维窗帘和油毡地板。

十二西大厦计划获得两个LEED白金认证,成为体现可持续设计尖端技术的实验室。它包括街头的零售空间、4层的ZGF办公空间、向外出租的17层"生态时尚"家园和5层停车场。这栋建筑还拥有生态屋顶、屋顶花园、露台空间以及完善的健身室和电影院。将4个可见的风力涡轮机组安装在高层建筑上,在美国尚属首例。十二西大厦不仅给为生活奔波的城市居民提供了心灵港湾,还是一个展现未来可持续发展建筑设计的示范项目。十二西大厦即将获得两个LEED白金认证——一个获奖项目是新建工程,另一则是ZGF办公室的室内商业设计。

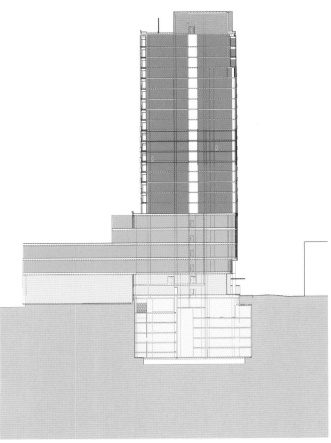

考虑到这栋建筑的较低楼层将用于商业活动，楼内所有房间不会被周边建筑遮挡并拥有全方位视野。273 靛蓝公寓起始于十二西大厦 6 层，提供 50~197 平方米规模不等的一间式、两间式、3 间式、工作室类型的住宅单元以及 3 层的复式住宅。

建筑模型预测——本案节约能耗的设计将减少 44% 以上的能源消耗，高于 2030 年此类项目的最高基准。提高效率的措施包括：蓄热体；日光及人体感应传感器；节约生活热水的低流量固定装置；高效率设备；热能回收设备；风扇辅助的夜间冲洗地面设备；办公楼层地板下的冷光束和配风设备；控制大空间通风量的二氧化碳传感器。

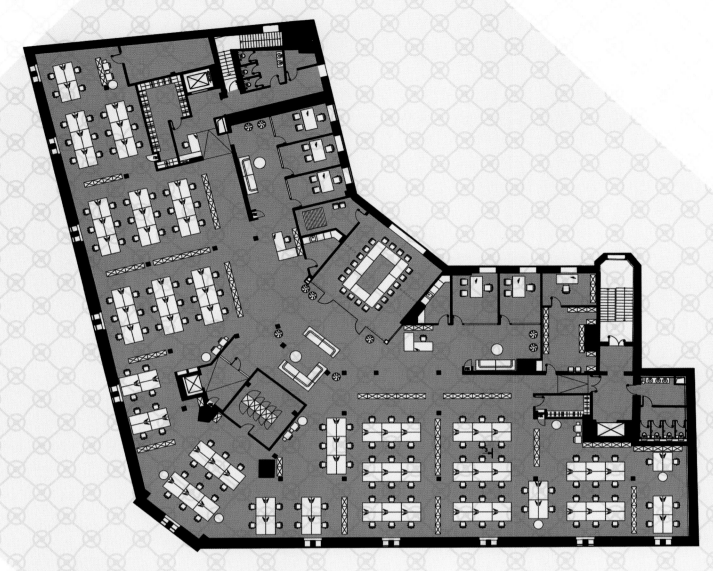

OFFICE OF BANK OF MOSCOW

莫斯科银行办公室

AUTHOR'S INTERIOR DESIGN
(АВТОРСКИЙ ДИЗАЙН ИНТЕРЬЕРА)

设 计 师：Alexey Kuzmin
建筑面积：7 000 平方米
项目地点：俄罗斯莫斯科市
摄　　影：Alexey Knyazev

Architect: Alexey Kuzmin
Area: 7,000 m²
Location: Moscow, Russia
Photography: Alexey Knyazev

The office of Bank of Moscow is located on two floors of a building which was built in the 19th century at the heart of Russian capital and situated at the intersection of two historical streets. The typical architecture, historicity and specific decorations of the building became inspiration of the designer of the first level of the office, where was planned to accommodate the reception and two departments of the bank. In decoration, designers used luxury wood panels with a pattern borrowed from the architectural details of the facade and marble. The very center of the floor is decorated with glass in the shape of a diamond. The second level of the bank is located in the attic—this premise has never been used, so that was done without the historical touch. The design of this part of the bank was developed as a simple, comfortable and modern office with the large open space for the staff, some training rooms and negotiation rooms, which can easily be transformed in size.

莫斯科银行的办公楼位于俄罗斯首都莫斯科的中心城区，它坐落在两个古老街道的交汇处，占据了一座19世纪建筑的其中两层。办公楼首层的设计者以典型结构、历史感和特有的装饰为灵感源泉，在此层设置了接待处和银行的两个部门。在装修上所采用的豪华实木地板借鉴了大理石建筑的外观细节。地板的中心则装饰有钻石形玻璃。该银行的第2层位于阁楼内，由于阁楼尚未被使用过，所以完全没有岁月的痕迹。银行区域的设计旨在体现简洁性、舒适感和现代化风格，并为员工提供充足的开放式空间和一些空间尺寸灵活的商务会议室、培训室，其大小可以任意调节。

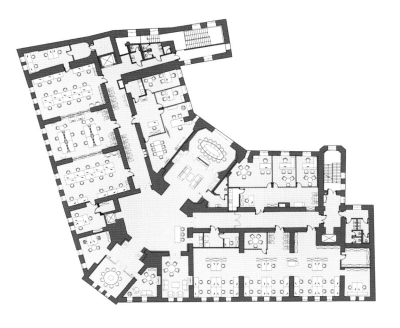

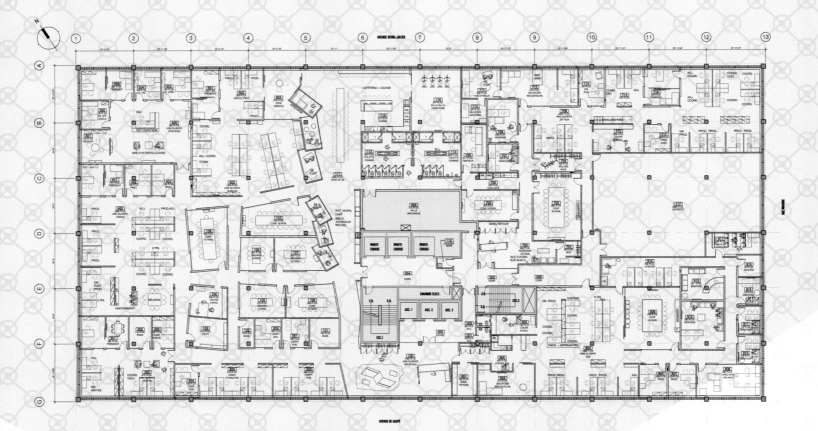

ATTRACTION MEDIA OFFICE

Attraction 媒体集团办公室

Sid Lee Architecture

建筑机构：Sid Lee Architecture
建筑面积：约 3 720 平方米
项目地点：加拿大魁北克省蒙特利尔市
摄　　影：Sid Lee

Architects: Sid Lee Architecture
Area: About 3,720 m²
Location: Montreal, Quebec, Canada
Photography: Sid Lee

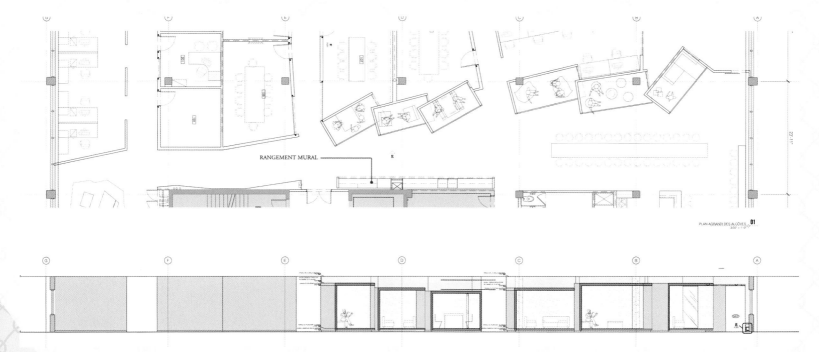

The project consisted of relocating the offices of Group Attraction Media and regrouping the following divisions of the company under one roof without dissipating their respective identities: Jet Films, Bubble Television, Cirrus Communications, Delphis Films, La Cavalerie and Attraction Media.
The project covers an area of 3,720 m². Inspired by urban landscapes, the chosen concept transforms the office into a small-scale city, completed with neighborhoods, plazas, streets and perspectives.
The various divisions of Group Attraction Media were transformed into so-called neighborhoods, each boasting its own design and color code. Every office is located near a large central space (reminiscent of a public plaza in the city). A bistro in the heart of this space welcomes employees working in the six surrounding offices. The design of each space was inspired by the activities of the division it houses. Finally, the spaces boast unobstructed views of the cityscape, the Olympic Stadium and the Jacques-Cartier Bridge. Our architectural challenge was to accommodate the various needs of the different companies with a limited budget and to structure a space of this size while respecting the personal identity of each company.

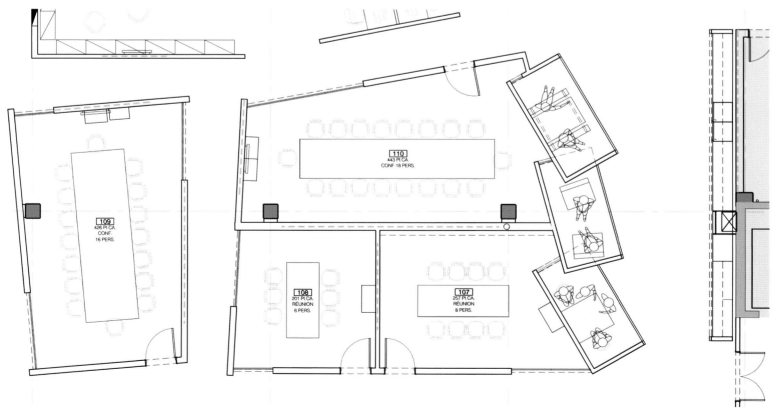

此项目包括两项内容，一是重新拟定 Attraction 媒体集团办公室的室内布局，二是在满足办公室设计的多样性需求的前提下重组几个集团下属分部的建筑结构，包括 Jet 制片、Bubble 剧组、Cirrus 通信、Delphis 制片、La Cavalry 公司和 Attraction 传媒。

该项目工程总面积达到 3 720 平方米。设计师从城市景观中获得灵感，将办公室打造成一个小型城区，并引入社区、广场、街道和远景等元素。

Attraction 媒体集团的各分部办公空间成为互动社区，并且各区域不同的颜色设计显示了空间需求的多样性。各个办公室均邻近一个宽广的中央区域（城市中公共广场的模拟形象）。这一区域中心设有一个咖啡馆，为周围 6 个办公区域的人员提供服务。办公空间根据各部门的主要工作活动内容进行划分。并且该设计使用户可以饱览附近的城市景观——奥林匹克体育场和雅克卡地亚桥等。此项目面临的最主要的设计挑战在于设计师需在有限的预算条件下满足各分部办公空间对个体特点和布局大小等方面的需求。

Cuisine avec îlot central permettant d'accommoder deux clientèles (traiteur et lunch).
Configuration à aire ouverte pour dynamiser l'aire de vie commune.

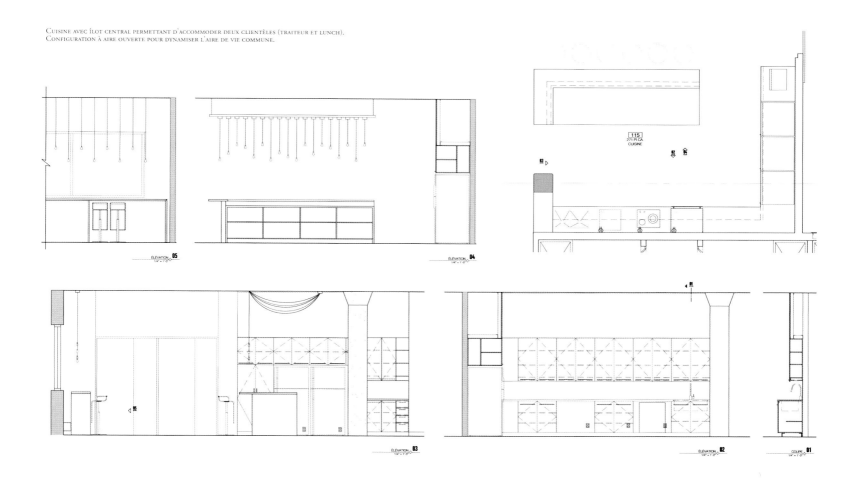

Rythme des murs et des portes.
Expression d'une verticalité et d'une séquence d'éléments dans l'espace.

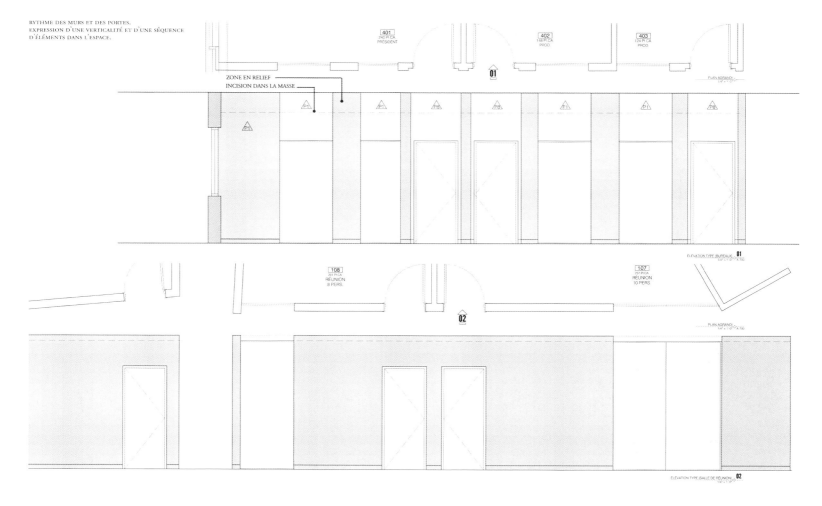

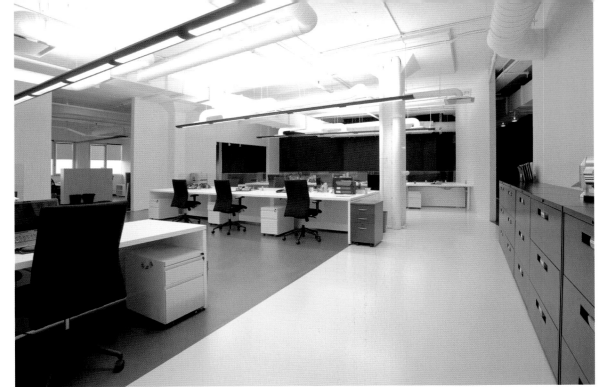
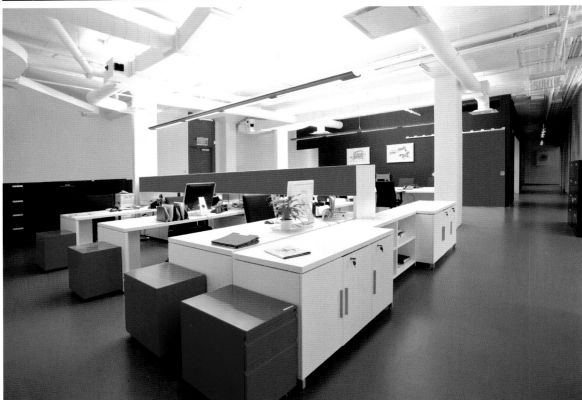

ENTRÉE SIGNALÉTIQUE D'ATTRACTION MÉDIA
EXPRESSION EN CONTINUITÉ AVEC LA PAROI URBAINE POUR UN PLUS GRAND IMPACT VISUEL.
INTÉGRATION D'UN GRAPHISME SUR LES PAROIS INTÉRIEURES.

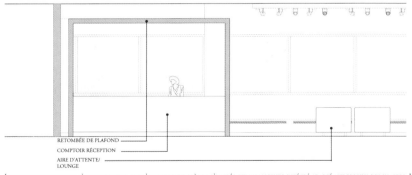
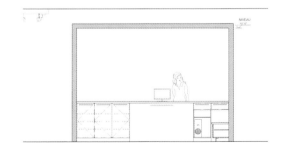

- RETOMBÉE DE PLAFOND
- COMPTOIR RÉCEPTION
- AIRE D'ATTENTE / LOUNGE

LE RANGEMENT MURAL DE L'ESPACE CENTRAL QUE L'ON RETROUVE PRÈS DE L'ENTRÉE EST UN MOBILIER INTÉGRÉ ADAPTÉ AUX BESOINS DES USAGERS. IL SE VEUT UN ESPACE DE SERVICE ESSENTIEL AU QUOTIDIEN. IL INTÈGRE L'ESPACE CAFÉ, LE PRÉSENTOIR DE JOURNAUX ET DE REVUES, UN ESPACE D'AFFICHAGE POUR LES COMMUNICATIONS INTERNES, DES ÉCRANS DE PROJECTIONS, DES ESPACES DE RANGEMENT DIVERS, ETC.

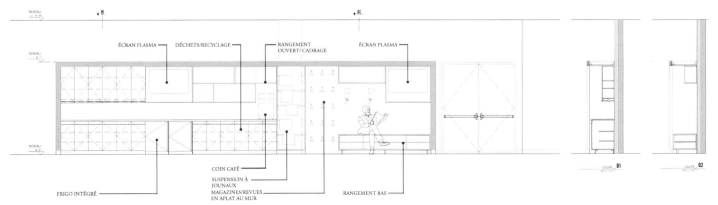

- ÉCRAN PLASMA
- DÉCHETS/RECYCLAGE
- RANGEMENT OUVERT/ CADRAGE
- ÉCRAN PLASMA
- COIN CAFÉ
- SUSPENSION À JOURNAUX MAGAZINES/REVUES EN APLAT AU MUR
- RANGEMENT BAS
- FRIGO INTÉGRÉ

PARSONS BRINCKERHOFF

柏诚集团

Incorp Design Studio

项目负责方：Incorp Project Management
建筑面积：5 400 平方米
项目地点：澳大利亚墨尔本市
客　　户：Parsons Brinckerhoff (PB)
摄　　影：Andrew Wuttke, Wuttke Photography

Project Manager: Incorp Project Management
Area: 5,400 m²
Location: Melbourne, Australia
Client: Parsons Brinckerhoff (PB)
Photography: Andrew Wuttke, Wuttke Photography

Parsons Brinckerhoff (PB) is one of the world's leading planning, environment and infrastructure organization. Their specialist consultancy teams deliver solutions for transport, renewable energy, urban development, water, and industry. PB and Incorp had been working together for many years, designing and delivering new workplaces in other Australian states before being engaged to design and project to manage their new Victorian headquarters.
The Melbourne office was a rabbit warren of organically grown spaces distributed over seven disjointed floors. This arrangement created a fractured workforce and reinforced silos. This outdated workplace was a legacy hangover from "old school" management thinking and hierarchies. Apart from the inconsistent messages this environment sent, the workplace was inefficient and uninspiring.
Despite the fact that the environment lagged behind other PB offices and more importantly competitors' environment, the PB culture was still collaborative and positive. As competition for expert staff grew, the pressure to lift their game with regard to team facilities increased. It became obvious with a lease expiry that it was time for a reinvigoration and re-engineering.

The Incorp team used its previous experience and knowledge of the PB culture and work practices to assist in the development of a comprehensive design and end user brief. The Melbourne team was consulted on many levels to flesh out the ultimate requirements for their future work place environment. Once that was finalized and the square meter footprint agreed, Incorp developed some building selection criteria to assist in the assessment of short listed buildings. Due diligence was then carried out on a very shortlist, including "test fit" planning to provide feedbacks on the most suitable location. This process leveraged the expertise of the entire Incorp suite of services including design, engineering and project management. PB then negotiated the lease arrangements to secure their space of 5,400 m², 3 floors at 28 Freshwater Place overlooking the river and the downtown CBD. Critical to the selection was the large floor plates of 1,800 m² with a side core allowing for open and highly flexible space. PB and Incorp were keen to ensure that the new design not only captured the corporate culture of the global organization but also catered for the Melbourne localization that would create a comfort level for both internal and external customers. While culturally the PB team is encouraged to be innovative, it was important that the design was not too radical but be in keeping with the relatively conservative nature of the client base and their expectation. PB has a long and successful history of designing and delivering landmark projects creatively and cost-effectively and needed this to be showcased. PB wanted to display its expertise and demonstrate that it was a safe and dependable choice to their potential clients. The client facing areas had to tread the fine line between conservative and dependable with innovative and creative. The public areas used a neutral palette with uncomplicated but high quality finishes to provide a classic and timeless air. The reception also subtly integrated the gallery of historic landmark projects into the walls to reinforce PB's track record of achievement. Incorp developed the designs closely with the PB executive team which included HR and marketing. This showed how PB understood the symbiotic relationship that the workplace has with its users and external customers. The visual design elements were carefully balanced against the brand values, budgets, time frames, technology requirements, functionality, workflow, team dynamics, flexibility, future proofing and the need to achieve a five-star Green Star accreditation. As environmental engineers, PB has the Green Star accreditation as an extremely high priority on its long list of project parameters. Behind the scenes in the "engine room", the general workplace environment has to be what the previous workplace was not. It has to be open, light filled, flexible, collaborative, highly functional, healthy, OH&S compliant, barrier and silo free, inspirational and energizing, ultimately creating a high-performance culture leading to greater productivity. All workstations are modular and standardized from a kit of parts allowing total flexibility. Workstation screens are kept low and all workpoints are within 8m from the windows, creating a light filled environment with sightlines across the entire floor. Greenery was integrated into the storage systems to soften the feel and to improve air quality and acoustics. Various styles of collaboration spaces were created to cater for the many types of meetings that occur both formally and informally. Meeting "pods" are used as colorful folies within the open plan to create interest and identity and also doubled as utility hubs. Scattered throughout the floors, individual quiet rooms provide a respite from the open plan and allow focused work or private calls. All solid fixed walls are kept to the internal zones away from the windows to allow maximum flexibility and openness. To "force" collaboration and interaction across the floors, PB provided a large central café that was designed to reflect the more urban feel of the local café culture. The new workplace has captured the organizations aspirations and manifests them into an open and flexible environment that inspires its users and encourages collaboration between both internal and external customers. The look and feel are in line with the global brand values but truly reflects the innovative yet slightly conservative culture of the local team while upholding PB's lofty environmental targets. This flexible space will grow and support PB's needs well into the future.

柏诚集团是一家在规划设计、环境与基础设施等方面处于世界领先地位的工程顾问公司。其专家级的顾问团队长期为交通、可再生能源、城市开发、水资源、工业等领域提供工程设计方案。Incorp 和柏诚集团有过多年的合作经历，在被委托设计柏诚公司维多利亚新总部之前，就参与了该公司在澳大利亚其他州的集团分部办公区设计项目。

原先的墨尔本办公室分散在 7 个楼层，由于不断增长的各种办公需求，使得其办公环境犹如一个杂乱不堪的养兔场。这种分布使员工之间的互动出现了分离和桎梏。这一办公空间是一种"旧式学校"管理理念下的朽败残余产物。除了员工互动不畅以外，原先的空间还影响了工作效率，不利于灵感的迸发和创新能力的提高。

这一办公环境不仅落后于柏诚其他分部，更重要的是落后于竞争对手。柏诚的企业文化一向重视协作，要求积极进取。随着专业员工数量的增加，提高人员团队素质、增加团队设施也迫在眉睫。很显然随着租约到期，原有的办公空间需要进行翻新设计。

Incorp 团队充分借鉴先前与柏诚集团合作的经验，结合对柏诚集团企业文化和工作实践的理解进行了综合设计。为了满足墨尔本分公司对于将来办公环境的最终需求，设计团队在很多层面上对他们进行了详细的咨询了解。Incorp 团队一旦明确这些需求并锁定好项目所在区域，就立刻制订出相应的设计参数以便对入围的建筑方案进行评估，拟定包括选址测试和反馈信息在内的候选地名单。这一过程充分发挥 Incorp 公司在设计、工程和项目管理方面的一系列专业技能和行业优势。后来经与柏诚集团协商，将这座占地 5 400 平方米的办公区新址定位在清水广场的 28 到 30 层，这里能俯瞰到河流和 CBD 商务区。这一选址的决定性因素在于该大楼每层面积达到 1 800 平方米并且其侧面是完全开放的灵活自由空间。

柏诚和 Incorp 共同关注的是确保新的设计既能彰显柏诚作为全球性组织的企业文化，又能反映墨尔本本土文化，创造出让国内外客户都感觉舒适且和谐的空间。柏诚的企业文化虽然注重倡导创新，但是在办公空间的设计上则要求根据员工的基本情况和期待留有一定的保守性，这也是设计时着重考虑的因素之一。柏诚集团拥有众多地标性工程设计的成功经历，其中所体现的创造性、预算使用高效性也得以在本案中体现。柏诚集团希望能完美地表现公司的行业专业性，并且向潜在客户展现他们的可靠性。因此设计师在进行客户接待区域的设计时，力求将保守性和创新性相结合，而公共区则是用一种简洁、高品质的中性色调来营造一种经典永恒的氛围。接待区域墙上也巧妙地悬挂展示了柏诚参与过的地标性工程的主题画，以此突出展示柏诚公司的成就履历。Incorp 在进行设计的过程中柏诚集团的人力资源部、市场部的执行团队保持着紧密联系。这种设计后的空间充分反映了柏诚集团对工作空间与员工、客户之间的关系的深刻理解。视觉设计元素和企业品牌价值、预算、时限、技术要求、功能性、工作流程、团队动态、流动性、未来适应性以及五星级绿星认证的目标等保持了微妙的平衡。作为环境工程师，柏诚一系列项目参数在赢得绿星认证方面具有极大的优势。

从"引擎机房"的内部场景中就可以看出总体的工作环境与原先截然不同。现在的办公空间更加开阔通透、结构灵活，易于开展协作且实用性高，满足职业安全健康的标准，整个空间自由畅通，充满创新灵感和活力，同时精美的设计文化也会激发高效的生产力。为了保证整体的便捷性能，所有的办公台选用标准的零部件，进行模块化组合。工作台的隔屏较低，每个工作点距离窗户不超过 8 米远，整个楼层都会被通透的自然光笼罩。融合进储备系统的绿植也很好地柔化视效，改善空气质量，降低噪声。风格各异的联合区间可以用来进行各种正式和非正式会议。会议室"荚"配色随意、设计形象独特有趣，彼此连接可形成实用枢纽。在地面上分散布局地设置一些小隔间，可以在其中集中注意力工作或接听私人电话。所有的立体固定墙均位于远离窗户的内部区域，以便能保证空间最大程度的灵活便捷、开放。为了加强内部协作和沟通，柏诚集团在中央位置特意设置了大型咖啡厅，这更加彰显了都市感和本土的咖啡文化。

新的工作空间能更好地增加团队凝聚力和认同感，开放灵活的环境可以激发员工的创新灵感，鼓励员工内部或者同客户间的沟通协作。设计的外观、氛围感觉都与国际品牌价值保持一致，真实地反映了本土团队在创意中略带保守性的文化，也彰显了柏诚集团推崇的环保理念。这一充满灵性的空间也必将很好地满足柏诚集团的发展需求。

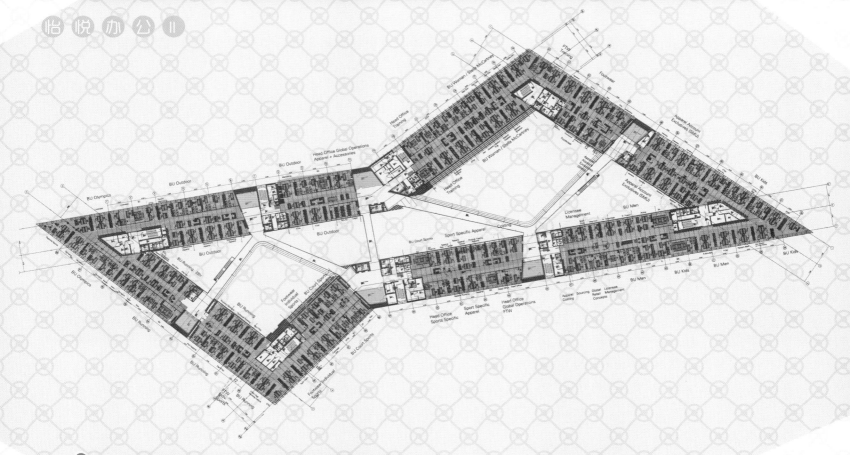

ADIDAS WORKOUT

Adidas 德国总部

KINZO

建筑机构：KINZO 建筑事务所
设 计 师：Martin Jacobs, Chris Middleton, Karim El-Ishmawi
建筑面积：61 900 平方米
项目地点：德国黑措根奥拉赫镇
摄　　影：© 沃纳 Huthmacher, © Lichtblick Fotografie Volker Bültmann

Architects: KINZO
Designers: Martin Jacobs, Chris Middleton, Karim El-Ishmawi
Area: 61,900 m²
Location: Herzogenaurach, Germany
Photography: © Werner Huthmacher, © Lichtblick Fotografie Volker Bültmann

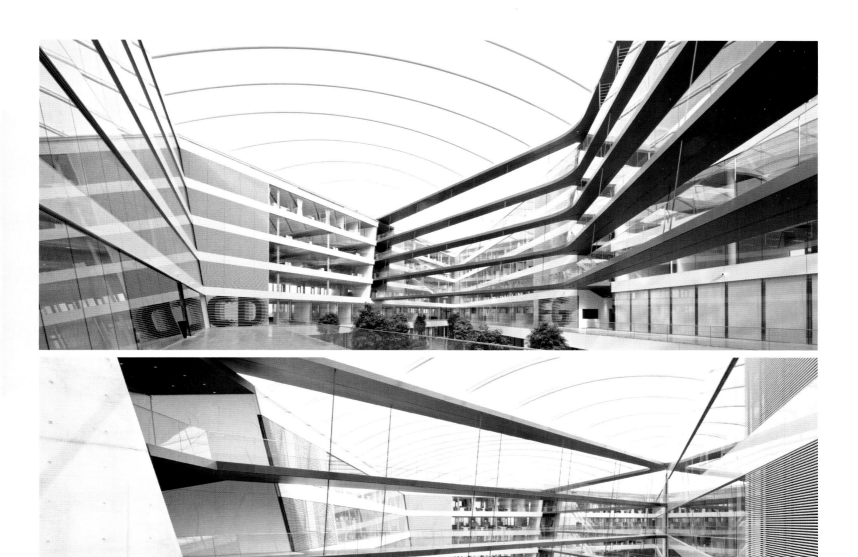

"WORKOUT" is the working title of the concept which is destined to open up new realms of creativity and possibilities for team play. To suit this purpose, KINZO developed a module system consisting of 46 flexible elements which can be put together in innumerable ways to establish team-oriented rooms, zones and workspaces. The "Teamplayer"—a multi-functional room module with conical sides—serves as the arrangement's centerpiece. Similar to an open frame, the "Teamplayer" can be equipped and used in various ways: As a support for desktops or as a clothes rack for lockable containers and magnetic punched-plate walls offer ample space for hanging up balls, shoes and assorted accessories. Drafts and photographs can easily be pinned to the walls with magnets.

And characteristic for every team effort, no element works entirely on its own—all parts are functional and esthetically designed to work together: Furniture works as a team.

The punch-plate elements are not only functional, they but also create an appealing optical effect—seen from a distance they almost appear to be transparent, allowing a "scenic" filtered view of the entire office floor. Experienced close up, however, they seem less transparent and offer the workspace a greater feeling of privacy. The "Teamplayer" in the central zones can be equipped with additional elements to convert them into storage spaces or "copy cubicles" for copy machines.

A small angle, a powerful effect: The trapeze-shaped side panels of the 1.43-meter-high pass on a dynamic, rhythmic structure to the open office floor. When the "Teamplayers" are placed on top of each other, striking wing-shaped partitions emerge, they give the room a stunning appearance. A team office, which is defined on two sides by Teamplayer elements, evokes the image of a honeycombed "spaceship" —in this way a room within a room is established, transparent and airy but also clearly staked off as the workspace of a particular team. Since every element has two sides, it can be individually equipped by two different teams, creating two separate workspaces, but it also visually connects workspaces in an overall spatial perspective. Seen from the central "corridor", the "Teamplayer" partitions convey more privacy than a conventional partitioning system arranged perpendicular to the windows. Contrary to a traditional partitioning system, it is possible to fine-tune every workspace with an individual arrangement or additional elements and to limit the "view" if desired.

> "WORKOUT"是这个设计概念的暂定名称,它旨在为团队工作开辟具有创造力和发展潜力的新空间。为了实现这一目的,KINZO开发了一个由46种灵活元素所组成的模块系统,它通过无数种组合方式来形成以团队为中心的房间、区域和办公场所。这种多功能模块化的"团队合作"模式配以锥形的边缘结构,成为策划的核心。这种"团队合作"的模式与开放式框架相类似,可以灵活应用于各种方式,如作为台式电脑的支撑、一个可上锁的密闭式衣架,或作为磁性墙壁穿孔板,为球类、鞋子和各类配件提供了充足的空间。如此一来,草稿和照片可以很容易地用磁铁固定到墙壁上。

整合众多特色设计为各个团队服务,任何一种元素都无法独立支撑整体运转——所有部件结合自身功能性与美学价值,协同发挥作用。家具元素也是团队协作中的一部分。

穿孔板元素不仅具备功能实用性,它们还创造了一种引人入胜的光学效应。从远处看,它们几近透明,可以将滤过的"风景"映射在整个办公室的地板上。但是走近的时候,透明会感弱化一些,穿孔板元素又为工作区域提供了更强的隐蔽感。中央区的"团队合作"模块通过配备附加元素,可以变为存储空间或放置复印机的"复印隔间"。

小幅度的倾斜角度产生了不同凡响的效果：1.43米高的吊架侧板为开放式办公地板提供了一个动态、富有节奏感的结构。当"团队合作"模式被运用于这些结构时，引人注目的双翼分区得以形成，为整个办公室换上了新面貌。也就是说，由于设计要素的双重性作用，团队办公室让人联想起蜂窝状"飞船"的形象。因此在办公空间内又设置了透明通风的独立空间，但也清晰地标明此空间专为特殊人员设立。每种设计元素都具有两面性，从局部来看它可以被两个不同的小组单独配备，创造两个分离的工作区域，但从整体的空间角度来看又具有视觉连贯性。从中央的"走廊"来看，"团队合作"的分区模式与传统的垂直分区系统相比更具有隐密性。与传统的分区系统相反，通过单独设置或附加元素补充，"团队合作"的分区模式可实现每个办公空间的细微调整和观景视野的随意调节。

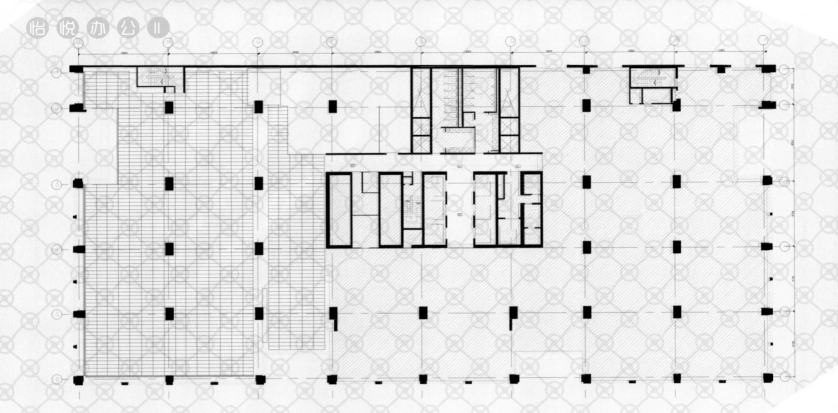

SINCLAIR KNIGHT MERZ PTY LTD.

Sinclair Knight Merz 公司

Incorp Design Studio

建筑机构：Incorp Design Studio
建筑面积：10 400 平方米
项目地点：澳大利亚墨尔本市
摄　　影：Anthony Fretwell

Architects: Incorp Design Studio
Area: 10,400 m²
Location: Melbourne, Australia
Photography: Anthony Fretwell

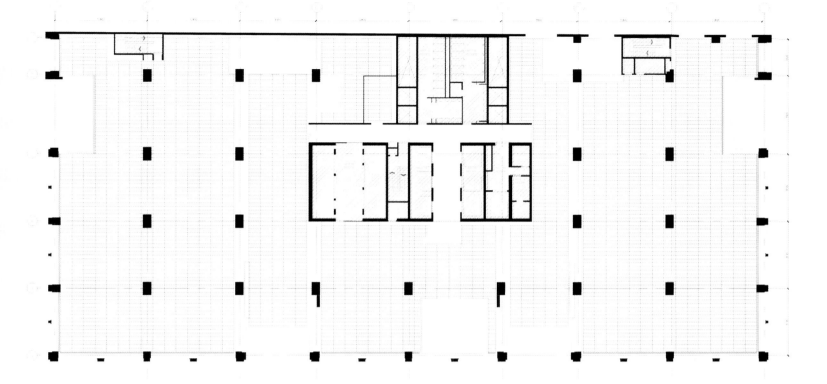

Sinclair Knight Merz (SKM) is a global provider of consulting services focused on the design and delivery of major construction and infrastructure projects around the world. Increasingly aware of their needs to continually innovate and lead their industry with environmentally sustainable solutions, they sought to reinvent and reinvigorate their new workplace. SKM's previous Victorian Headquarters was tired and outdated. In short, their workplace did not reflect who they were and who they aspired to be.

With this in mind, SKM engaged in the services of Incorps' Interior Design Studio to work closely with SKM's project and design management team, to develop the end user brief and the concept and detail design of the new environment. SKM's property team had already negotiated the new space in Melbourne's CBD consisting of 10,400 m² existing office space. The project had several risk factors to be solved, including incredibly tight timeframes and the fact that the space needed completely to be stripped out and made good priority to commencement. Added to this, the need was for the integration of an interconnecting stair over 4 floors and the absolute requirement to achieve a five-star Green Star accreditation.

Conceptually the new workplace had to clearly signify to the market that SKM was the innovative industry leader. The new space had to embody the cultural values publicly espoused by SKM which included environmental sustainability, innovation, responsibility, democracy, collaboration, inclusiveness, progressiveness, flexibility, modesty, respect and expertise. SKM's new direction was to enable a full integration of its services seamlessly across desks, offices, floors, cities, states and countries. They have created "ONE SKM" and the new Victorian headquarters was to be reflective of that vision.

SKM's organisation is built on expert individuals working in fully integrated teams that ebb and flow with project and client needs. As such, the new workplace needed to be incredibly flexible to support those business dynamics, allowing collaboration in many forms, being formally planned or serendipitous. For this reason the large floor plates were kept as open as possible with as little fixed elements as possible. Traffic flows were clearly delineated by floor finishes and lighting clues, to ensure the least disruption to focused workers. A variety of workplace settings were made available to accommodate the various preferences for individual work and team work. The concept of the "nest" was pursued as a metaphor for a safe, nurturing and organic home. This resulted in very natural and neutral tones for the base palette, imbuing the space with a warm, friendly and serene feeling. Access to natural light was critical in the layout of the floors, necessitating the open plan workstations within 8 m of the windows and all meeting facilities in bound. The flat hierarchy and inclusiveness was taken literally to exclude all individual offices. There was to be no barriers to collaboration and communication. Overlaying the natural base palette was pockets of color and vibrancy that were used as collective hubs to refresh and reinvigorate. These colorful and energetic breakout spaces and cafes were distributed throughout the floors to allow immediate access for brainstorming. Further, creative integration of company values and vision was used in graphic solutions throughout the space. Technology hubs were minimized to reduce SKM's carbon footprint, but remote technology access was stepped up to allow collaboration and knowledge sharing/access to happen anywhere and any time. The desire for integration and accessibility across the businesses manifested in the multi floor interconnecting stair that was showcased at the entry zone, providing greater transparency for SKM's partners. The top floor provides training rooms and breakout spaces that flow to an open roof terrace that is accessible to all staff. Every finish and item specified were rigorously tested against the environmental requirements to ensure the least energy, resource consumption and complete Green Star compliance.

The new workplace designed by Incorp, truly captured the cultural aspirations and corporate values of SKM. It reflects what they stand for and where they want to go, ensuring the current and future team members "walk the talk", and are culturally aligned. SKM's new high-performance workplace provides them with a competitive edge for the attraction and retention of the highest calibre team. The ONE SKM "nest" will support them wherever their aspirations and vision take them.

SKM是专业从事全球性设计、提供重大基础设施建设工程咨询服务的公司。SKM逐渐意识到企业需要不断创新、走可持续性的环保产业道路,公司力图重整现有的工作环境。SKM前维多利亚区总部的格局显得颓败过时,概括地讲,他们现有的办公空间既不能体现企业现况,又无法体现其发展目标与前景。

因此,SKM委托Incorp室内设计工作室协同SKM项目管理团队,共同拟定出终端客户需求概要和新办公空间的具体设计方案。SKM公司资产管理团队已将新的办公地址定在墨尔本中央商务区内一处总面积达10 400平方米的空间里。本项目亟待解决的问题是在动工前极其有限的时间内拆除原有的布局设计,改用互联楼梯连接4个楼层,完成符合五星级绿星认证标准的设计方案。

理论上新的办公室设计应该对外彰显出SKM的创新型企业形象,符合SKM长期信奉宣扬的价值观念,包括环保、创新、责任、民主、协作、包容、先进性、灵活性、谦虚、尊重与专业。SKM新的发展方向是跨越桌子、办公室、楼层、城市、州或国家的界限,打造全方位服务。他们借由设计维多利亚区办公总部实现打造"独一无二的SKM"的愿景。

SKM的组织形式是专家个人和完全磨合好的团队协作以满足和跟进工程项目和客户的需要。因此,新的办公室必须具有极大的自由度才能更好地开展各种形式的团队协作,无论是正式的还是偶然的。因此在这个大的空间范围尽可能安排比较少的固件,以使空间看起来尽可能地开放通透。过道区用不同的地板饰面和光线流清晰界定出来,用以减少对集中精力工作的人造成的干扰。多种工作区间形式可以满足无论是个人工作还是团队活动的各种需求。"巢"的概念其实是对一个安全、培育性、有机的家的一种隐喻表达。因此设计者采用自然和中立的基调,使空间沉浸在一种温馨美好、宁静安详的氛围中。地表布局是影响自然光线倾透的关键所在,在开放的计划下工作台应该距离窗户8米以内,所有个人办公区间不分层级,极具包容性,因此交流和协作也变得毫无阻碍。叠加的天然基色是色彩和活力的口袋,可以作为枢纽用来更新振兴公司。色彩丰富活力无限的休息室和咖啡厅遍布整个楼面,随时可以商议沟通集思广益。

另外,设计师将公司价值观和愿景创意式地融入到图形之中并通过空间展示了出来。技术中心被最小化以减少办公室的碳足迹,但是远程技术被引进以便沟通协作,信息知识传递能够不受时间和地点限制地进行。对外合作上的联合和准入的需求特点在入口处多层互联楼梯的设计中完美地体现了出来,为SKM的合作客户提供一个方便了解自身的透明开放视角。顶层空间的原有布局被更改用做培训室,以及一个员工们都可以休息驻足的开放性露台。每一项物件和饰面都根据环境要求进行了严格的测试,以确保降低资源消耗来实现绿色之星认证。

这一新的办公室由 Incorp 一手打造,很好地把握并展现了 SKM 的企业精神和价值观。体现了他们的立足点和未来愿景,确保现在和将来的团队成员自由交流合作并且充满凝聚力。SKM 这一新的精品办公空间带给他们一种可以吸引和保留优秀团队成员的竞争优势。这个独一无二的 SKM "巢"形办公室将会一直支持着他们一展宏图。

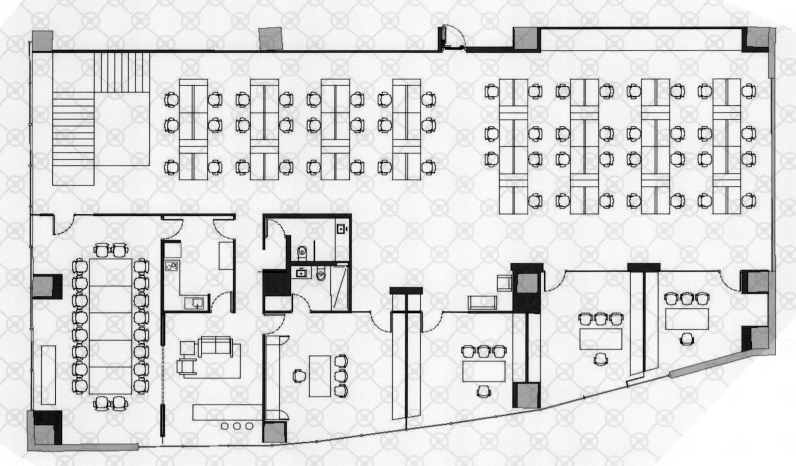

PAGA TODO OFFICES
Paga Todo 公司办公室

Usoarquitectura

设 计 师：Gabriel Salazar, Rnando Castañón
建筑面积：2 000 平方米
项目地点：墨西哥 墨西哥城
摄　　影：Héctor Armando Herrera

Architects: Gabriel Salazar, Rnando Castañón
Area: 2,000 m²
Location: Mexico City, Mexico
Photography: Héctor Armando Herrera

Space is the main factor that determines the interior design and operation of a company. The new corporate offices for Paga Todo presented a particular challenge because it was necessary to adapt to the clients' demands of a 2,000 m² area in a shopping center.

A big wood box, inserted respecting the surrounding design, greets everyday collaborators and visitors. Inside the box were located the reception, support area and interview halls, on top of it—with a panoramic view of the finance area—the personalized area to serve the dealers.

The client decided to implement in their office a lounge style cafeteria—like a hotel lobby— because before the relocation the majority of the collaborators preferred to work and meet in the close by cafeterias to enjoy a more relaxed ambiance. This space has all the necessary services and it is a nice surprise for the visitors because there is screen, complimentary computers with Internet, snacks and drinks. The staircase was located in the vertex of the project in order to communicate with the upper level, opening a new entrance of light from above and making more interesting this meeting point for the colleagues.

The color palette—asked by the client is very sober and with no risk—white, beige shades with accents in a dry green and the oak of the furniture and woodwork. Three sections with meeting halls divide the space, generating references and transitions between the work cells.

For natural light big vertical stripes were open on the facade of the shopping center and most of the walls were not built to ceiling height to make the most of the different natural light sources of the building. Large windows facing the interior of the shopping center were also installed to make references in the main corridor. The windows have random size and create a sequence with the transition of each of the work teams.

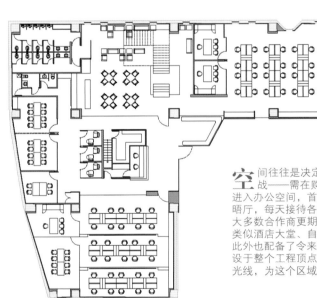

空间往往是决定室内设计和公司运作的一个主要因素。Paga Todo 新办公空间的设计面临了一个突出的挑战——需在购物中心 2 000 平方米的空间里打造一个各方面均符合客户需求的空间形象。

进入办公空间,首先映入眼帘的是与周围设计风格相辅相成的大型木质包厢。包厢内设置接待区、附属区和会晤厅,每天接待各路来宾。包厢上层区域可以一览金融区全景,区域内设有可与商户进行个人化会面的场所。大多数合作商更期待可以在类似自助餐厅的轻松氛围中工作与会谈,因此在重新拟选新办公室地址时,确定了类似酒店大堂、自助餐厅休息室的轻松设计风格。这一新办公空间配备所有公司正常运作所必须的设施元素,此外也配备了令来宾耳目一新的设施,如显示屏幕、免费联网电脑、零食小吃和饮料等。

设于整个工程顶点的楼梯方便不同楼层人员进行沟通交流。楼梯与顶端相接之处留有透光口,方便摄入充足的光线,为这个区域里的员工们带来不少趣味。

空间的基调按客户的需要选用了非常素淡平和的白色和米色,在干绿色橡木家具和木制品中显得别具一格。三个会议厅将整个空间隔开,起着工作区之间的过渡指引作用。

为了增加自然采光，购物中心的立面采用大规模的竖条式开放性设计，多数墙面采用落地窗，以便大楼可以尽可能地吸收各种自然光源。在购物中心大楼的主廊道上也安装有朝向室内打开的大型窗户。窗户的尺寸随机设置，借以区分每个团队的工作区域。

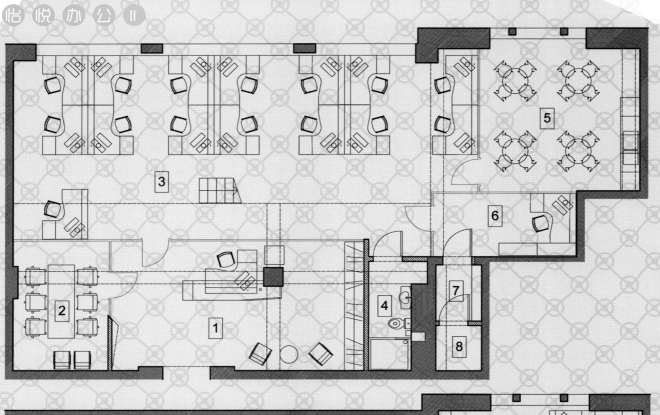

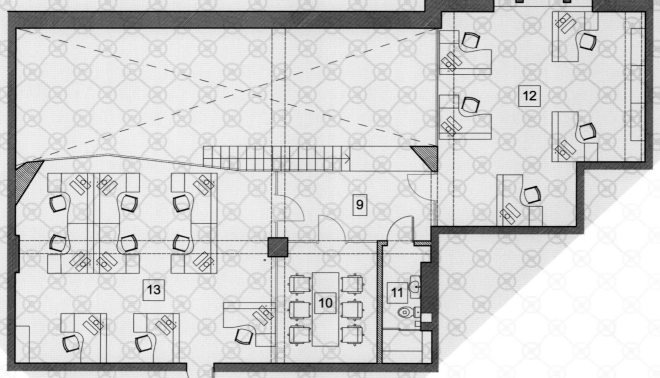

YANDEX KIEV OFFICE

Yandex 网络公司基辅办公室

Za Bor Architects

项目团队：Peter Zaytsev, Arseniy Borisenko
建筑面积：290 平方米
项目地点：乌克兰基辅市
摄　　影：Peter Zaytsev

Project Team: Peter Zaytsev, Arseniy Borisenko
Area: 290 m²
Location: Kiev, Ukraine
Photography: Peter Zaytsev

The Yandex Company, for which this office is designed, is the most popular Russian Internet segment search engine with lots of useful services, ranking in world-top 25 sites. The company is constantly developing, so in late July, the Yandex Kiev branch moves to the new office. As a dozen other Yandex offices, the design is still in charge of the Moscow bureau's Za Bor Architects.

Arseniy Borisenko and Peter Zaytsev—the architects are making comments on their new design: "Every project we strive is to make the most eco-friendly."

It appears in the use of mostly natural materials and creation of user-friendly spaces: Comfortable, quirky, even cheerful spaces do not resemble dull gray offices in their standard plastic version. While developing the concept of the Yandex Kiev Office, designers expected to create a modern and welcoming space—that is, infuse the high quality of this IT-company, which is famous for the careful attitude of their staff.

There were no special problems in designing and negotiating since the first office developed for Yandex. This office volume has been allocated a tiny 290 m² (the smallest office designed by the bureau today), and the location is very interesting. Future office has to take the sixth and seventh floors of A-class business center "Leonardo", which was a single two-level space with a void. The room is well lit with natural daylight through a huge arched window which offers a wonderful view of the Kiev opera house and the area in front of it.

While in contrast to other Yandex offices, there is no gym and other recreational zones except for a rather big coffee-point, it is very convenient and not overloaded with workspaces. There are about fifteen workplaces equipped with ergonomic Herman Miller furniture on each of the two levels. Visitors are met at the original reception area where they can comfortably wait for a meeting in contemporary Fritz Hansen armchairs. The reception area is designed in corporate colors and symbols.

On each of the two levels there is a meeting room. The bottom one is used to accept clients, the upper one to hold staff meetings. The top level is praiseworthy, which is fully glazed and tied up to the bottom with the staircase. The project was planned to have an extremely complex geometry of the gallery and staircase, which should behave as an entity. To explicate the contractors its layout, the architects produced a model of the staircase block. In late winter, the model was brought from Moscow to Kiev. Thanks to builders in Kiev—they are managed well with an exact fit of planes and glass gallery, so that the design looks exactly as planned. After completion, the architects took the object builders' report that this was the most difficult job that they had to perform for their practice.

这个办公室是为俄罗斯最受欢迎的多功能搜索引擎门户、全球互联网排名前 25 的 Yandex 公司设计的。由于公司规模不断壮大，Yandex 基辅分公司于七月下旬搬进了新的办公室。和其他一系列 Yandex 办公室一样，这个新办公室仍然交由莫斯科的 Za Bor 建筑事务所设计。项目设计师 Arseniy Borisenko 和 Peter Zaytsev 这样评价自己的作品："对于每项工程我们都会追求最环保的设计。"

正如评价一样，设计中采用最纯粹的自然材料创造出以人为本的办公空间。这个舒适、新奇且令人愉悦的场所完全不同于其他标准塑化的沉闷灰暗空间。在体现发展Yandex网络公司基辅办公室的设计理念的同时，设计师试图打造一个现代化且受欢迎的空间，也就是说，在设计中融入IT公司典型的认真执着的工作态度和价值观念。鉴于设计师多次为Yandex公司设计办公室，在设计和沟通方面并无难处。该办公室所选空间只有290平方米，是现有Yandex办公室中面积最小的，不过其所在位置颇为有趣。新办公室位于甲级商务中心利奥纳多大楼的第6和第7层，这里原是一座有着架空区的独立双层空间。此空间配有一个大拱形窗户，因此自然采光条件良好。凭窗远眺，基辅大剧院及剧院前面的美妙景观一览无余。

这里与其他Yandex办公室相比，仅配有一个相对较大的咖啡厅，并无健身房和其他娱乐休闲区。但是此设计简洁便利，也不会给工作区造成干扰和负担。在这两层的办公空间里一共分为15个工作区，均配有符合人体工学的Herman Miller高档办公家具。在入口不远的创意接待区里配有当代的Fritz Hansen扶手椅，以便来宾可以舒适地等候交流会面。接待区的规划设计同时采用了Yandex公司典型的色调和标志。

在这个两层空间里，每一层均配有一个会议室。下层的会议室用于客户洽谈，上层的会议室常用于企业内部的员工会议。上层办公室值得称道的是它采用全玻璃透明设计，利用楼梯连通下层办公室。该项目设计中楼梯和长廊的几何形状极其复杂，需利用实体模型向承包商阐明其布局。建筑师制作了楼梯模型，并在深冬季节将模型从莫斯科带到基辅。基辅的施工者们根据模型精准地处理了平面和玻璃画廊的结合安装问题，因此完工后的设计和原规划误差极小。项目竣工后，施工者向设计师表示，这个是他们开展施工实践以来难度最大的项目。

KNOCK INC.

KNOCK Inc 公司

Julie Snow Architects Inc.

建筑机构：Julie Snow 建筑设计公司
建筑面积：930 平方米
项目地点：美国明尼苏达州明尼阿波利斯市
摄　　影：Paul Crosby

Architects: Julie Snow Architects Inc.
Area: 930 m²
Location: Minneapolis, MN, USA
Photography: Paul Crosby

KNOCK Inc. is home to a branding, advertising and design firm whose growth is based on a collaborative business model that brings together a diversity of design disciplines to serve their clients. KNOCK was in search of either a building or neighborhood within the city that could serve to feed this creative culture and become their new headquarters.

The project began with the purchase of a neglected 1960s office building located along Glenwood Avenue just north of downtown Minneapolis. We were asked to re-envision, renovate, expand, and create a new KNOCK workspace that could enhance their collaborative work model, convey their unique brand and connect the buildings and interiors with Glenwood Avenue and the unique mix of residential and industrial building stock that surrounds the neighborhood.

The renovated space and new addition create a 930 m² workspace and a distinguished new presence enlivening Glenwood Avenue.

The existing building had "good bones" but was in desperate need of repair and updating. As first time building owners, KNOCK required low maintenance, sustainable solutions with maximum performance. In order to achieve this goal, our sustainable strategies were focused on passive design solutions that offered maximum benefits with minimal initial costs. High-performance insulated walls, roofs and glazing replaced single pane glass units, non-insulated exterior walls and a failing roof. This was supplemented with heat recovery units that transferred heating and cooling from exhaust air to fresh air intake. To eliminate the need for artificial light during the day and enhance the work environment, all working areas of the facility have access to daylight through new and expanded window openings along the perimeter of the building and the extensive use of skylights and sola tube light collectors along the interior core. Healthy and local building materials were chosen such as reclaimed walnut for the main wood pieces of the building and cedar siding for the additions at the front and back of the facility. Most of the existing structures are exposed, minimizing excessive ceiling materials and increasing the sense of openness. The site is within one of Minneapolis' Pedestrian Overlay Districts sandwiched between a light industrial area and residential neighborhood and thus requiring close collaboration with the city in all sites and building design issues. To facilitate the building's interaction with its location, KNOCK's activities are integrated with the surrounding community. Located along Glenwood Avenue, a large dynamic workspace framed in glass connects interior activity with that of the street.

A cedar wood volume penetrates the transparent facade to form a new, wood-lined entry conference space, branding the building with KNOCK's signature wood grain pattern. A more private cedar deck relates to the residential neighborhood to the south and offers dramatic skyline views. Native vegetation buffers the building and the surrounding site while minimizing long-term maintenance.

KNOCK's creative and collaborative business model sets out to redefine what it means to build a brand. Their office space was designed to support this idea and enhance teamwork and creative interaction by using gathering and collaborative spaces as the organizing elements of the design, rather than placing them in a peripheral, or secondary, relationship to the individual work spaces. A diversity of collaborative spaces was created throughout the office, including pinning up walls in all work areas for spontaneous meetings, a focused critique space for small team meetings located in the heart of the office, a library stocked with the latest design-focused books and periodicals for smaller and quieter gatherings, and two large conference rooms for office and client meetings. The collective team culture is reinforced in the kitchen with a table designed for meetings, casual gatherings and lunches of twenty plus employees. A yoga room, exercise room and a game room provide additional non-traditional work areas. The work, the culture, and the workspace combine to convey the very unique creative product offered by KNOCK.

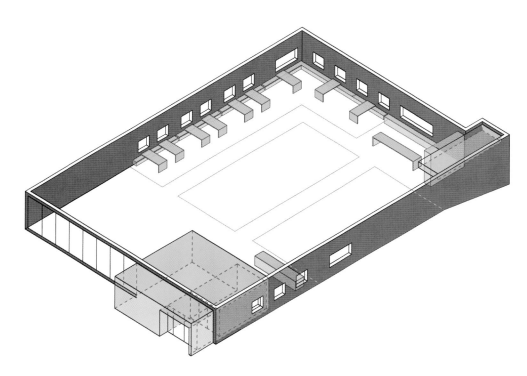

KNOCK公司始终坚持综合发展模式，涉及品牌、广告和设计领域，旨在为客户提供多样化设计服务支持。KNOCK公司希望在原办公室所在建筑或者市内附近地带创建一个展现公司创新文化的新办公总部。

该项目位于明尼阿波利斯市中心北部的格伦伍德大道旁，一座建于上世纪60年代不起眼的办公楼内。该大楼被买下后，设计师受委托将其重新设计、扩建并改造成一个新的办公空间。客户要求新的设计须促进协作这一工作模式并传达独特的品牌观念，使室内设计与格伦伍德大道上其他建筑及其周围独特的商住混合型住宅相融合。

经过设计改造，这一930平方米的办公空间使整条格伦伍德大道焕然一新。

原有建筑架构完善，但仍然需要进行大规模的维护和翻新工程。KNOCK公司在项目之初就计划将对原有建筑的改造最小化，以可持续的设计方案来最大程度地满足功能需求。为了实现这一目标，设计师探寻可行性策略，力求投入最小成本以获得最大效益。故采用高性能隔热墙、屋顶和复合玻璃取代原有的单层组合玻璃、非隔热外墙和年久失修的屋顶，并且辅以热量回收机组，经过制冷或加热排出废气而引入新鲜空气。为了减少白天人工照明的使用以及改善办公室环境，扩大了空间四周原有窗户的尺寸，打通了新的天窗，在室内核心地区装置太阳能集光管，使所有工作区域设施能充分利用自然光。设计中充分利用当地环保型建筑材料，比如在主要建筑木器件上采用的再生胡桃木，在其他结构正反面使用的雪松壁板。设计师将大部分建筑结构外露，尽量减少屋顶上繁杂的装饰设计，以增加空间的开放性。项目位于明尼阿波利斯市轻工业区和住宅区之间的人流密集地带，因此不论在地址选择还是建筑设计方面都要求与城市有着密切的联系。为了与建筑所在地区形成互动，KNOCK公司的设计规划与周围社区环境相融合。沿着格伦伍德大道，在办公室的内部与临近街道处采用玻璃相隔的设计，形成了与城市充分互动的办公空间。由雪松木组成的室内空间穿插于透明的玻璃立面，形成由木质线条勾勒的新会议空间。建筑上的LOGO设计同样采用木纹样式，使得公司品牌更加醒目。一个凸出的私人雪松木露台与南部居民区相连，形成美妙的天际线景观。原生植物缓和了建筑和周围生态的冲突，且无需长期维护。

KNOCK创新和协作的商业模式阐明了其品牌意义。整个设计组织聚合多个空间，而不是简单地将其他空间放在单一办公空间的附属关系上，更好地体现并且强调了团队协作和创意互动的品牌理念。整个办公室由一系列的多样联合空间组成，包括由可装订墙体组成的突发会议办公区、办公区中心内可进行集中讨论的小型会议区、为安静的小型聚会准备且藏有最新设计类书刊的图书室、两个可进行客户洽谈和员工会议的大型会议室。厨房设有长桌，可容纳20多名员工会谈、休闲聚会、用餐，进一步深化了公司的团队合作文化。另外这一办公空间内设有健身房、瑜伽房和游戏室，使其明显有别于传统办公空间。工作、文化和生活空间的完美结合形式也可以称作是KNOCK公司的一项独特的创意产品。

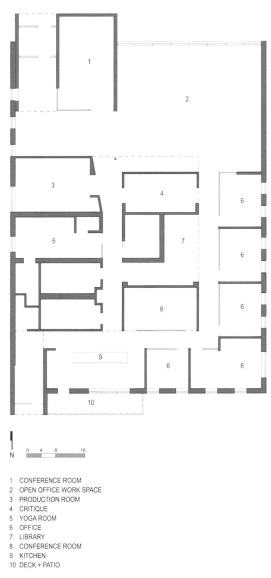

1 CONFERENCE ROOM
2 OPEN OFFICE WORK SPACE
3 PRODUCTION ROOM
4 CRITIQUE
5 YOGA ROOM
6 OFFICE
7 LIBRARY
8 CONFERENCE ROOM
9 KITCHEN
10 DECK + PATIO

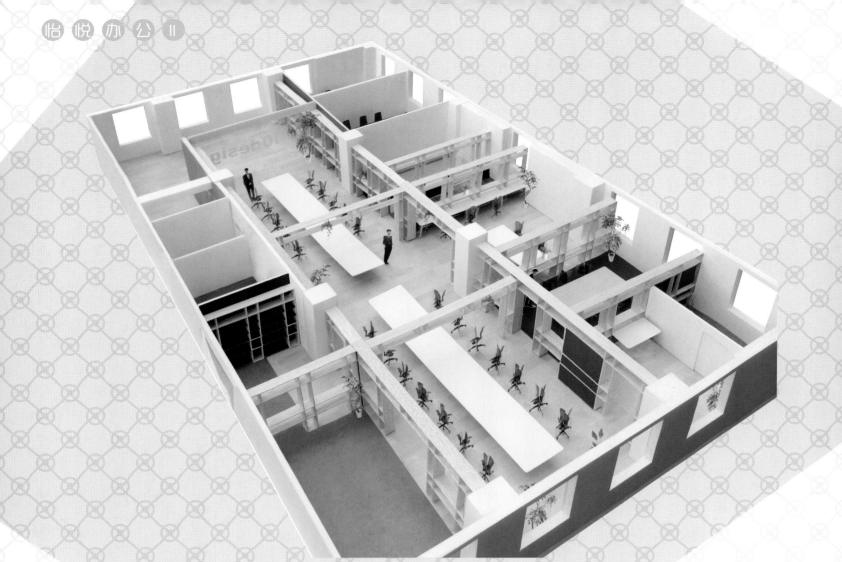

1-10DESIGN KYOTO OFFICE
1-10design 网页设计公司京都办公室

Torafu Architects

设 计 师：Koichi Suzuno, Shinya Kamuro, Eikichi Saku
建筑面积：462.57 平方米
项目地点：日本京都市
摄　　影：Daici Ano

Architects: Koichi Suzuno, Shinya Kamuro, Eikichi Saku
Area: 462.57 m²
Location: Kyoto, Japan
Photography: Daici Ano

The offices of Kyoto-based web production company, 1-10design (one to ten design), comprise a single 462.57 m² versatile area featuring work spaces, a meeting room, a gallery and a traditional Japanese room serving as a resting area.
These offices not only fit for business, but also for a variety of purposes. We proposed a spacious floor finely partitioned by wooden frames, allowing rooms of various sizes to flow into each other. The framework can also serve as shelves on which we can find office furnishings alongside personal belongings. The frames keep functional spaces separated as the rooms' appearance changes, and yet, they offer a peering view that gives the whole floor a sense of continuity.
The main work space straddles many rooms, but a long table running across the center connects them together. Adjacent to this work space is a meeting room, a laboratory, the president's office, a traditional Japanese room and a gallery.
Also, the wooden frames offer flexible storage combinations and a variety of usages by turning into book shelves, storage space or even a bench according to location.
This bare skeleton can become a wall as much as a piece of furniture but really serves as a background prompting miscellaneous interactions by keeping the ability to accommodate future changes with flexibility.
The office allows one to change the scene in front of us according to our mood while preserving the sense of unity of a single room.

1-10design 公司主营网页设计，其新办公室位于京都，总面积 462.57 平方米的办公空间包括一个多功能办公区、一间会议室、一个画廊和一个传统的日式休息室。
除办公室外，这一设计更加注重空间的多功能应用性。Torafu 建筑事务所采用大面积的木质地板结构，并以木框架分隔出空间功能区。木框架本身又可以作为置物架，用来存储私人及办公用品。木框架可以随意变换形式，也可以增设面板而形成封闭空间，这种方式不仅能灵活地应对办公室未来发展的需求，还能保持良好的整体空间连续性。

主要办公区域跨越了多个原有房间,以一张长办公桌连接。与办公区相接的是会议室、实验室、社长办公室、一个和式休息室和一个画廊。
木质框架除了作为储存空间外,还充当了灵活的书架、储物柜,根据位置情况还可以用作座椅。
这个空的搁架可以作为一堵隔墙,也可以是一件家具,但更多的是充当一种背景,它促进了办公室内部的工作与互动,同时可以适应未来空间灵活转变的需求。
这个办公室呈现出其灵活多变性的同时又能保持单一空间的整体性。

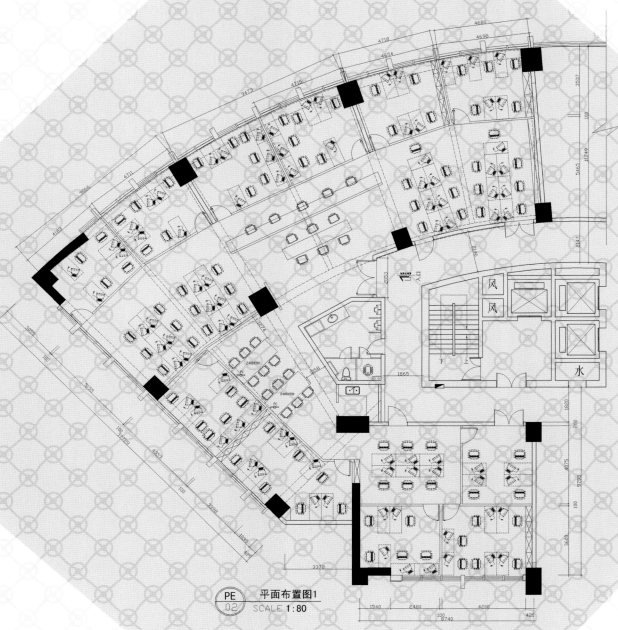

GUOGUANG YIYE XIAMEN BRANCH

国广一叶厦门设计分公司

Fujian Guoguang Yiye Building Decoration Design Engineering Co., Ltd.

设 计 师：叶斌
建筑面积：800 平方米
项目地点：中国福建省厦门市
主要材料：槽钢，青石，红木，玻璃

Architect: Bin Ye
Area: 800 m²
Location: Xiamen, Fujian, China
Main Materials: Channel Steel, Bluestone, Wood, Glass

The Guoguang Yiye Branch design is the perfect combination with Eastern and Western culture, combination of classical and modern, integration of simple, elegant, classic, fashion together, and people in the office can enjoy the surrounding scenery, which is the Guoguang Yiye Branch Design in Xiamen Office. This office is located in the CBD commercial center of Xiamen, but the high-pace work can not be invaded. This unique and elegant office adheres to the laid-back and quiet standing over the city.

Channel steel, bluestone, wood, glass and other materials are used personality full of different areas. Chinese and Japanese, East and West, classical and modern are perfectly combined. This office is not only a communication of fashion and classical, but also the recreation of the gas into the formal office club. Superb design gives new life to classical, elegant environment, the human hustle and bustle of city temporarily separated. In a noisy environment we can enjoy a trace of quiet, empty space in the work release tension.

Red, black, white of these rich Chinese pre-Qin culture, the ceiling decorated with Chinese paper-cut style of personalized lighting, together with Chinese classical home, combined with modern technology and European large area of windows, mirrors and day-style furnishings, from them, just like swimming between classical and modern, East and West. Different from the traditional-style office compartments, the case abandons the small space, replacing with the relatively wide space; people feel agreeable, when the office is also, less depression and tension, more than a hint of harmony and communication.

Abandoning the complexity of the sculpture, the case is replaced with clean simple lines of the furnishings. Put the art of calligraphy in the design so that the rigid space smarts up. Right lighting endows the concise space with demure beauty, from ceiling windows facing the road people can enjoy the beautiful and bustling city night. A small amount of green space embellishment adds dynamic, kind of comfortable and relaxed feeling.

JOY OFFICE DESIGN 2

将设计与东西方文化完美融合，古典中散发着现代的气息，融简洁、优雅、古典、时尚于一身，让人在办公之余尽情享受周边的美景，这就是国广一叶厦门设计分公司办公会所。它位于厦门中心商务区内，却丝毫没有受到繁忙工作节奏的影响。会所依旧以其独有的典雅悠闲形象，静静伫立在城市中。

槽钢、青石、红木、玻璃等个性十足的材质被用于不同的区域，将中式和日式、东方和西方、古典与现代这几种艺术风格完美融合。这种设计不仅为时尚的办公会所增添了历史韵味，还将休闲欢快的气氛融入到正式的办公环境中。高超的设计手法赋予了古典优雅的环境新的生机，将人们与城市的喧嚣暂时分隔开来。人们得以在喧闹的环境中享受一丝宁静，在空旷的空间里释放工作的压力。

以红、黑、白这些富含中国先秦文化的色彩为主色调，天花板装饰以中国剪纸式样的个性化灯饰配以中国古典的家居，再结合现代技术与欧式大面积的落地窗、镜子以及日式家居。当你置身其中，宛若遨游在古典和现代、东方和西方之间。与传统的隔间式办公室不同，本案摈弃了狭小的办公区间，以心旷神怡的开阔空间代之，这样在工作之时便少了几分压抑与紧张，多了些许融洽与交流。

本案以简洁的线条和简单的陈设取代了以往复杂的雕塑作品。将书法艺术融入设计之中，使刻板的空间结构增添了几分灵气。恰到好处的灯光运用使简洁的空间布局流露出娴静气息，人们通过面向公路的落地窗户可以尽情欣赏城市繁华秀丽的夜景。少量绿色植物的点缀增加了空间动态感，让人颇有一种怡然自得的轻松感。

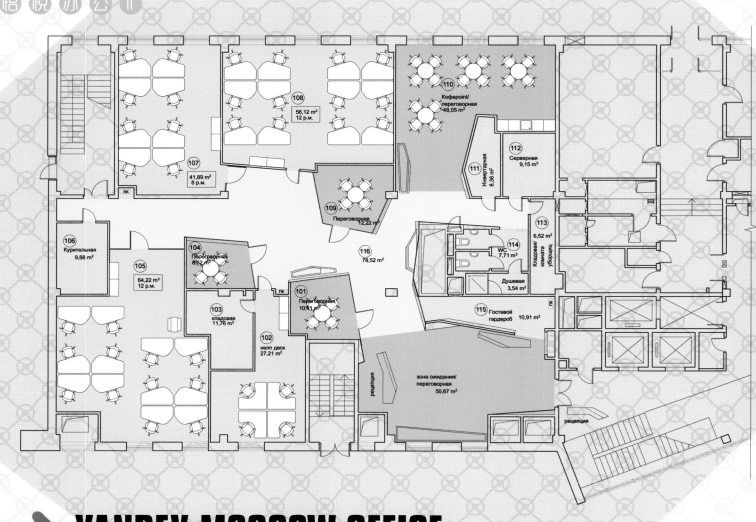

YANDEX MOSCOW OFFICE

Yandex 网络公司莫斯科办公室

Za Bor Architects

设 计 师：Arseniy Borisenko, Peter Zaytsev
建筑面积：7 000 平方米
项目地点：俄罗斯莫斯科市
摄　　影：Peter Zaytsev

Architects: Arseniy Borisenko, Peter Zaytsev
Area: 7,000 m²
Location: Moscow, Russia
Photography: Peter Zaytsev

Yandex is the biggest and leading Internet company in Russia, CIS and Russian-speaking countries. Yandex office in Moscow is the second office by Za Bor Architects designed for this company. The office occupies all seven floors in one of two wings in the newly constructed business center "Krasnaya Roza" (eng: Red Rose) in Moscow. The wing is close to rectangle in section and consequently the office space is concentrated around the technical core with riser pipes. This construction peculiarity makes it necessary to create sanitary facilities and other premises that demand water use—such as kitchens and coffee points—around them.
The original columns and work frames have been preserved. Considering the client's request, Za Bor Architects uses a layout with individual rooms strung along the corridors. To soften the impression, the role of internal walls is played by a glazed partition. Translucent orange coating has been used to add bright and vivid color to the gray and white interior. The ceilings have been visually expanded in the corridor, and the communications have been painted in deep gray color, while in the work areas ceilings from Ecophon, sound absorbing materials used in cinema interiors, have been used for additional sound isolation.
Internet wiring and electrical cables are made on the raised floor. As the flooring has been chosen to carpet tile, which allows you quickly access any point in the hidden under-floor communications.
Dynamic volumes and expressive furniture are the main hallmarks of Za Bor Architects style. This distinctive approach has been used by Za Bor Architects to develop the concept reflecting the idea of a sky-rocketed exponential Yandex development. Yandex offices are known for their informal atmosphere and attitude to working process. Since they work round the clock, besides large and small conference halls and traditional general working zones, the project includes a sports hall, a kitchen, coffee points, etc.

Yandex 是俄罗斯、独联体国家和一些俄语国家中最大的主流互联网门户。Yandex 莫斯科办公室是 Za Bor 建筑事务所为 Yandex 公司设计的第 2 个办公室。这个办公室占据了莫斯科新建成的"红玫瑰"商务中心两座 7 层裙楼中的一整座。办公室所在的这座裙楼靠近矩形结构的相接部分，因此这个办公空间被集中在由升流管组成的技术核心区里。针对这种建筑结构需专门建造卫生设施和其他日常所需的基础供水设施，比如为周围的厨房、咖啡厅等。
布局上，设计师保留了建筑原有的框架，将不同的房间沿着走廊连接起来，满足了客户提出的要求。透明的玻璃内墙使整个空间变得柔和，橙色涂料的使用则为黑白灰的整体风格增添了一丝明快感。走廊的天花板采用视觉扩大化的设计手法，会议交流区用以深灰色，工作区的天花板用的是 Ecophon，这种用于电影院内部设计中的吸声材料，可以达到更好的隔音效果。

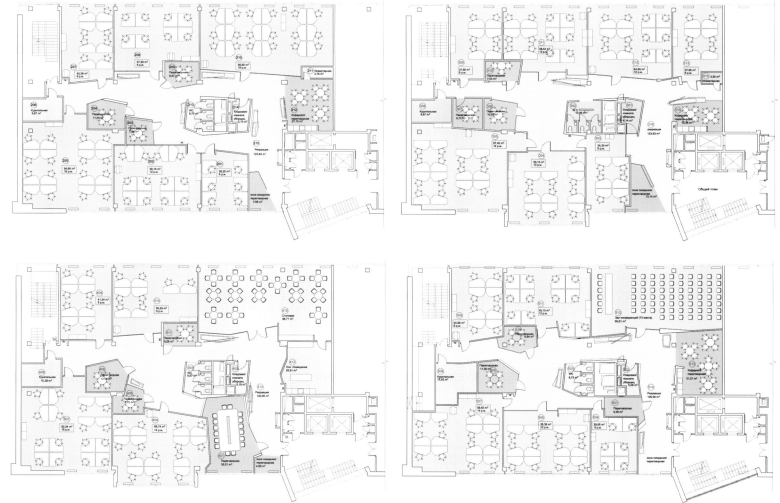

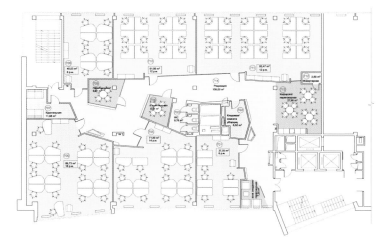

因为地板是砖质材质，这样置于活动地板上的网线和电缆便可以随意连接到埋在地板下的通讯交流设施。
富有变化的空间造型和极具表现力的家具是 Za Bor 建筑事务所的主要风格特征。他们已经系统化地运用这种独特的风格反映 Yandex 飞速发展。Yandex 系列办公空间因轻松写意的工作氛围而彰显魅力。Za Bor 设计团队夜以继日地完成了大小会议厅、传统大众工作区以及此项目的其他设施——健身房、厨房、咖啡厅等。

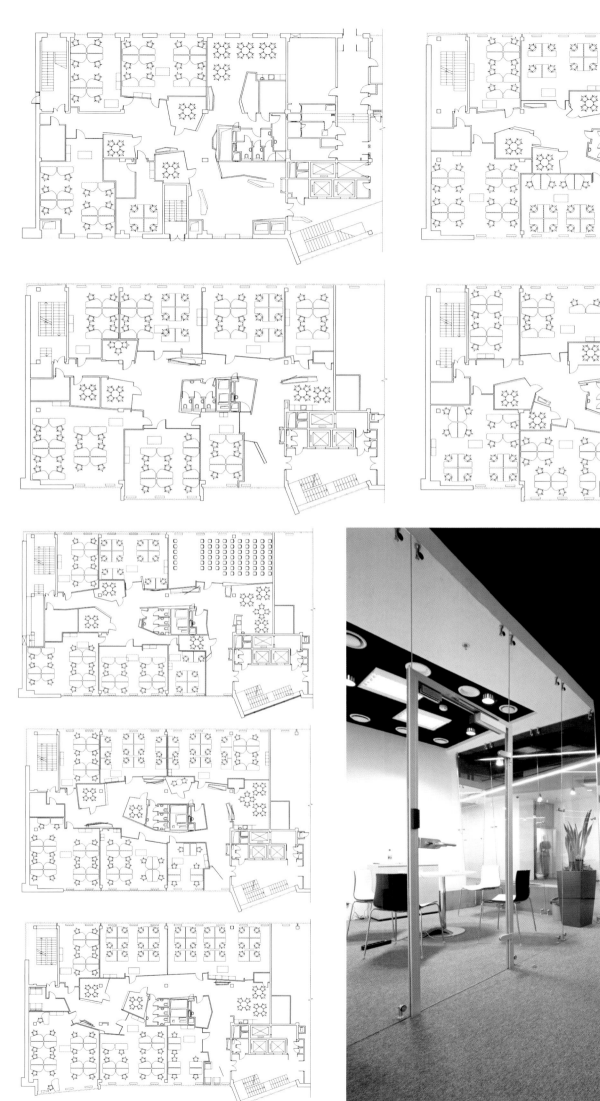

FORWARD MEDIA GROUP PUBLISHING HOUSE OFFICE

前锋传媒集团出版社办公室

Za Bor Architects

设 计 师：Arseniy Borisenko, Peter Zaytsev
建筑面积：4 200 平方米
项目地点：俄罗斯莫斯科市
摄　　影：Peter Zaytsev

Architects: Arseniy Borisenko, Peter Zaytsev
Area: 4, 200 m²
Location: Moscow, Russia
Photography: Peter Zaytsev

In addition to several popular and well-recognized magazines such as *Hello*, for instance, Forward Media Group, owned by Russian billionaire Oleg Deripaska, publishes the largest and the most popular magazines on interior design in Russia and Russian-speaking countries, including the main edition *Interior+design* and specialized periodical *100% Office*, *100% Bathrooms*, *100% Kitchens* and so on. Za Bor Architects was chosen among several thousands of architects to design Forward Media Group Publishing House Office. Office space was destined to be quite complex—it was a huge loft of 4,200 m², significantly elongated and located in the mansard level of a new business center. The situation was made worse by the peculiarity of the publishing house—the need for editorial offices with open spaces as well as commercial and retail departments, separated offices for directors and editors-in-chief with a conference corner, conference halls, archive with library, storage rooms, etc. All of these facilities were placed along the corridor going through the whole room. Eventually, open space offices were concentrated on the one side, and cabinets were mainly located on the other side of the corridor.

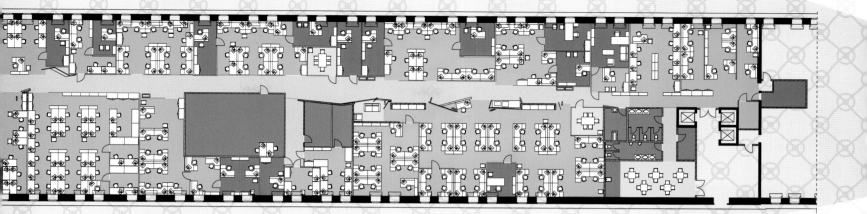

Communication points are supposed to be transparent and not to be vanished in a homogeneous office. As a result, the elevator lobby with stones, bright receptions, and meeting rooms is emphasized; a yellow complex construction hides the entrances to the toilets; archive room is marked by the black floral pattern. The same concern is the specific build-in furniture designated by Za Bor Architects for the visitors. In contrast, operational areas are designed in a neutral gray.
Technical communications are placed under a ceiling of the mansard level, as well as rafters and baulks have not been hidden, but painted in black instead, which visually elevated and extended the ceiling.
Recently the publishing house has moved to another building—historical Trechgornaya Manufacture building complex, but all the elaborate construction elements have been preserved and reinstalled in the new office.

俄罗斯亿万富翁奥列格·德里帕斯卡所创的前锋传媒集团出版社，除了出版了诸如《Hello》这样颇具名气的流行杂志以外，还出版了在俄罗斯和俄语国家十分畅销的室内设计杂志，其中包括《室内+设计》和专业期刊《100%完美办公室》、《100%完美浴室》、《100%完美厨房》等。Za Bor 建筑事务所从千名事务所中脱颖而出成为前锋传媒集团出版社的设计单位。
办公空间的设计注定是相当复杂的——这座 4 200 平方米的阁楼体态狭长，位于一家新商业中心区的双重斜坡屋顶楼层上。这样的地势特点结合出版社自身特定的需求使得设计任务更加艰巨——编辑部和商业零售部门的空间需呈开放式，主管与主编需享有独立的会晤空间，除此之外还应配备会议厅、图书馆档案室、仓库等。所有设施均沿走廊分布，贯穿整个空间。最终，开放式办公空间与独立办公室各自分布在走廊的两侧。
通信服务应是公开透明的，在同类性质的办公空间也应该坚持此原则。正因如此，石质电梯门厅、明亮的接待处和会议室均为设计中的重点；洗手间采用外部的黄色复杂构架隐藏入口；档案室选用黑色花卉作为标志性图案。同时 Za Bor 建筑师为访客特别选购嵌入式家具，而将运行区设计得相对中性化，并以灰色作为基调。各种科技通信设施位于双重斜天花板下面，和屋顶的椽结构、承载物一起不加掩蔽而涂以黑色，在视觉上提升和延长了天花板的层次感。
近期出版社搬进了历史悠久的 Trechgornaya 生产综合楼，但所有精致的房屋结构件仍被保留并在新办公室中重新得以运用。

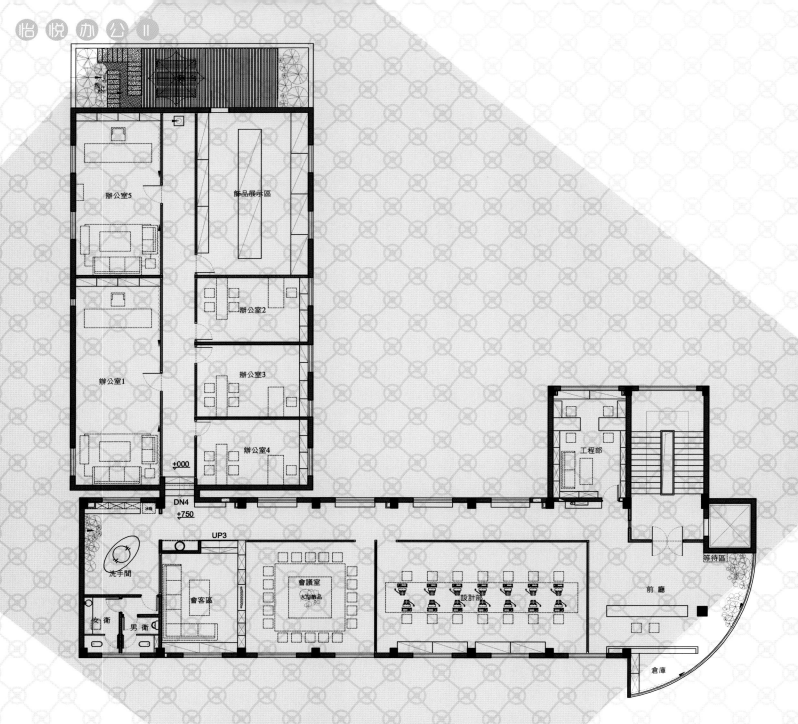

ZHENCHUAN SHOP BY XUPIN DESIGN LTD.

叙品设计震川店

Xupin Design Decoration Engineering Co., LTD.

设 计 师：蒋国兴
建筑面积：1 000 平方米
项目地点：中国江苏省昆山市
主要材料：蘑菇石，地毯，木条，墙纸，复合地板，玻璃，空心砖，白色乳胶漆

Architect: Guoxing Jiang
Area: 1,000 m²
Location: Kunshan, Jiangsu, China
Main Materials: Mushroom Rock, Carpet, Wooden Sticks, Wallpaper, Flooring,
　　　　　　　　Glass, Hollow Brick, White Latex Paint

Xupin Design was launched in a nice winter. In the view of many people, winter in south of China is moist and gray. But in our eyes, exactly because of the dull sky of winter, we are more conscious on the design works of people's life. The happiness of life comes from our believeness on wishes and imagination. With such a hope, we start our journey of dream.

In the color and layout of space, we run out from traditional way and lie our uniqueness in the using of white color. A large scale of white makes a scene and ambience of "music expressed then far less than silence revealed".

We borrowed rise after restrain way from the design of traditional gardens. The plain door, narrow corridor and hidden entrance are all changed into another world when you are turning around suddenly. When walking through the corridor lined with cobblestone, with a breeze caressing the cheeks, all the tiredness will be forgotten in that moment.

Green is a symbol of vigorous, while white is elegant. No matter hard outfit to furnishings or colors are harmonious and reflect designer's pursuits on details. Then people and objects here are in one.

Attributed to the theme of dream, Chinese culture here is performed greatly in a modern way. The space will always offer you unexpected and pleasant experience on designer's meticulous work. The briefness and purity theme also runs through the whole case.

Office is a place where people are engaged in mental work. The mood of employees, productivity is often varied in the environment. While in the new office designed by Xupin, with all of pleasant colors, unique ideas and tea aroma in the air, staff here can enjoy a more attractive environment and improve work efficiency.

This is not only a helped style of work, but also an expression of positive attitude towards their life. In such a materialistic age, the project is designed to arouse the naivety from the trivialities, and then to experience a more enjoyable and real life.

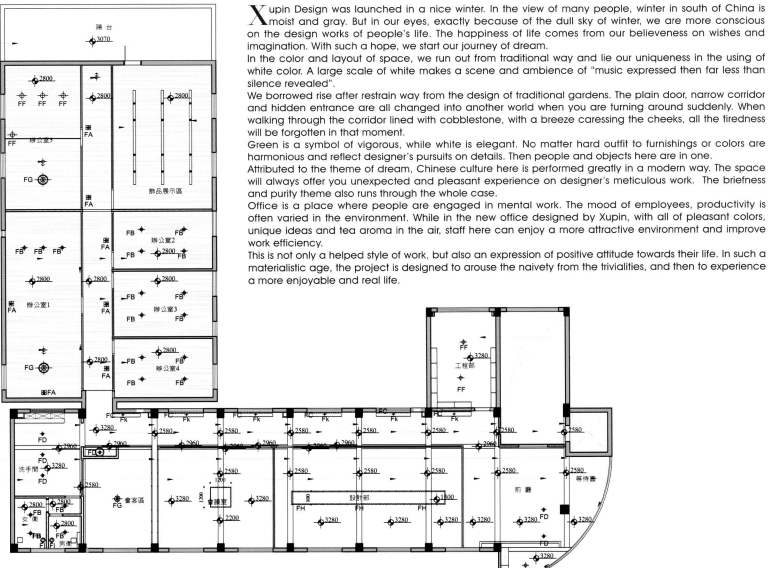

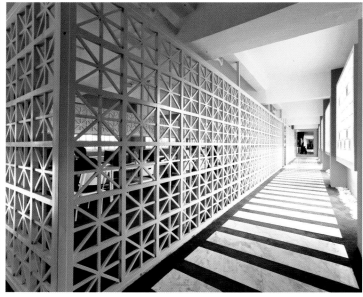

叙在 品诞生于一个美丽的冬天。在许多人看来,南方的冬天潮湿而灰暗;在我们设计师看来,因为阴沉,才更要装扮人们的生活空间;生活之所以美丽,是因为人们拥有美好的愿望和无穷的想象力,抱着这样的愿望,我们开始了梦想征程……

设计空间的色彩和布局上跳脱传统、独树一帜,大量运用白色元素。巧妙的白色运用以及其他颜色的搭配极具意境,营造了"此时无声胜有声"的氛围。

借鉴了传统园林设计中欲扬先抑的手法,低调的门、狭窄的走廊、隐蔽的入口都在转身的一刹那豁然开朗,别有洞天。鹅卵石夹道,古风荡漾却清新怡人。走过长廊,尘世的烦恼也慢慢抛诸脑后。

绿色象征着生命,白色象征着优雅。从硬装、家具陈设到配色都协调统一,体现了对细节的完美追求。人和物在这里共存,融为一体。

"梦幻"的主旨对中国文化进行了很好的现代化诠释表现,空间设计总会在细致入微的地方给你意想不到的美妙感受,简洁纯净的主题始终贯穿整个项目。

办公室是人们从事脑力劳动的场所,员工的情绪、工作效率常常会受到环境的影响。而在叙品设计的这间新办公室中,轻松愉快的色彩、别致巧妙的创意,再加上空气中弥漫的茶香味,都让工作人员在放松的心情下完成工作,从而促进了工作效率。

这不仅是一种健康的工作方式,更体现了设计者们积极的生活态度。在这物欲横流的年代,本案的设计希望使内心烦躁茫然的人们重新找寻到最初的那份质朴,在静下心来快乐工作的同时来感受真正的生活。

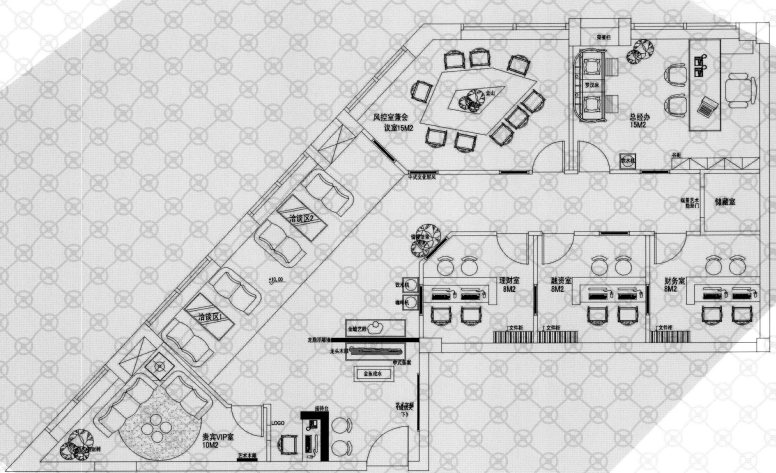

DRAGON · TRIPOD

龙·鼎

Henan Dongsen Decoration Engineering Co., LTD.

设 计 师：刘燃
建筑面积：180平方米
项目地点：中国河南省濮阳市
摄　　影：徐朝亮

Architect: Ran Liu
Area: 180 m²
Location: Puyang, Henan, China
Photography: Chaoliang Xu

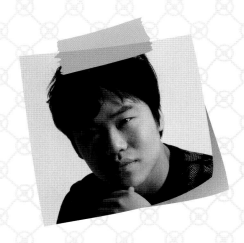

The case is about a financial investment company.
The designers link the space planning, design techniques and elements together with the subject closely and accurately and adopt low carbon and environmental-friendly materials (such as grass weaving lamp, root, steel net) to break the formative office space, which reflects a cultural space of Chinese feelings.
The dragon symbol at the vestibule is in the form of tripod and forms an organic whole with the Chinese long narrow table and the wooden carved dragon, which is the soul of this case. The suspensory S lamp (woven by grass) and the roots hanging on the wall add a relaxed, natural and harmonious atmosphere for the discussion area. The open irregular "golden hill block" in the conference room echoes well with the world to archive extremely strong visual effect and is in accordance with the nature of the company. The Chinese screen becomes the space carrier, linking the various functional areas together closely and reasonably.

这是一个关于金融投资公司的设计方案。
设计师在空间规划、设计手法及元素搭配上紧扣主题，并采用低碳环保的材质（如草织灯、树根、钢网），打破了格式化的办公空间，体现出具有中国情怀的文化空间。
玄关处标志性的飞龙符号以鼎的形式出现，并与中式条案、木质雕龙形成一个有机的整体，展现了本案的灵魂本质。墙顶悬吊的S形灯具（由草编织）与树根为洽谈区增添了轻松、自然、和谐的氛围。会议室不规则的开放式"黄金区域"极富时代感地展现强烈的视觉效果，与公司整体风格相呼应。中式屏风作为空间的承载者，把各功能区域紧密合理地联系在一起。

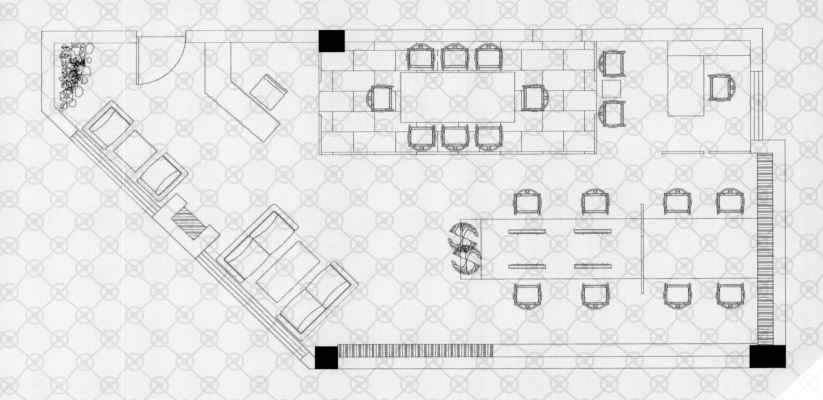

ELEMENT · PLAIN COLOR SPACE
素色空间

Henan Dongsen Decoration Engineering Co., Ltd.

设 计 师：刘燃
建筑面积：110 平方米
项目地点：中国河南省濮阳市
主要材料：质感涂料，木质花格，仿古砖

Architect: Ran Liu
Area: 110 m²
Location: Puyang, Henan, China
Main Materials: Quality Paint, Wooden Lattice, Pseudo-classic Brick

The case is about a decorative design company. The building area covers only 110 m². The designers meet the needs of integrating functions of reception, discussion area, conference area, office area, general manager office and other functions in the limited space by combining with the nature of the decoration company itself.
In the overall design, the designers pursue function first, nature and harmony, conciseness and fluency, integrate the Chinese elements into the space with the Chinese lattice as the soul, which links each functional area together organically. The space's color tone is based on black, white and grey, which reflects the simple, natural and cultural space connotation.

本案为装饰设计公司的项目。建筑面积仅110平方米，设计师在有限的空间内结合装饰公司本身的性质满足了前台接待、洽谈区、会议区、办公区、总经办等多种功能需求。
在整体设计上追求功能第一、自然和谐、简洁流畅，并把中国元素符号融入到空间表现中。中式花格是空间的灵魂所在，它把各功能区域有机地联系在一起。空间以黑白灰为色彩基调，体现出朴实、自然的文化空间内涵。

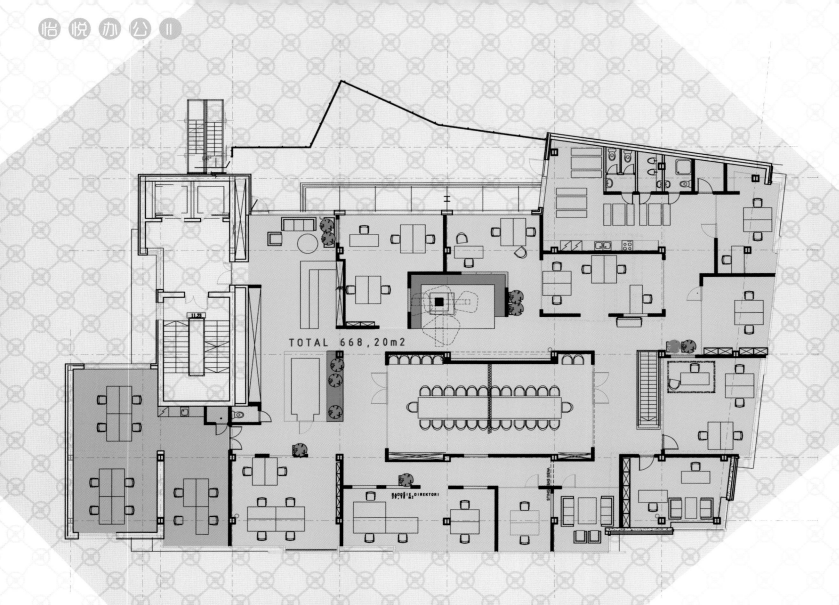

MCCANN-ERICKSON RIGA AND INSPIRED OFFICE

麦肯埃里克森里加创意办公室

Open AD

设 计 师：Zane Tetere, Elina Tetere
建筑面积：1 020 平方米
项目地点：拉脱维亚里加市

Architects: Zane Tetere, Elina Tetere
Area: 1,020 m²
Location: Riga, Latvia

Functional planning is made as stylization of union of inner room and outside—there are formed "streets" on which as separated blocks are located "houses" with windows, facades, podiums, in floor recessed flowerbeds, etc. For finishing and furniture is used simple materials—boards, lacquered MDF, white wall painting on 4th floor, grey lacquered blastula for 3dr floor walls. Interior is created in topical recycling style—4th floor wall and secretary panels are made from timber products remainders and no more usable household or office appliances. Lamps are made of high diameter plasterboard tubes which previously were used as rolls on which fabric and textiles are reeled. Low-budget projects worked with simple materials and developed "trash style" in interior projects.

该项目的功能规划目标是将内部的房间与外部空间以极富风格化的手法相连，从而在已形成的"通道"上建造配有窗户、立面、平台、凹型苗圃等的隔间。考虑到装修加工与家具的效果，木板、喷漆纤维板、4楼的白色喷漆以3楼墙壁的灰色喷漆毛胚都采用了简单的材料。室内设计是典型的循环利用特征，4楼的墙面面板及写字台面板均用剩余的木器及家庭或办公室的回收用具制作。台灯是以先前用作多种纺织品卷筒的高直径石膏板管做成。基于低预算的标准，建筑师运用简单材料打造了一个蕴含"废弃风格"的室内空间。

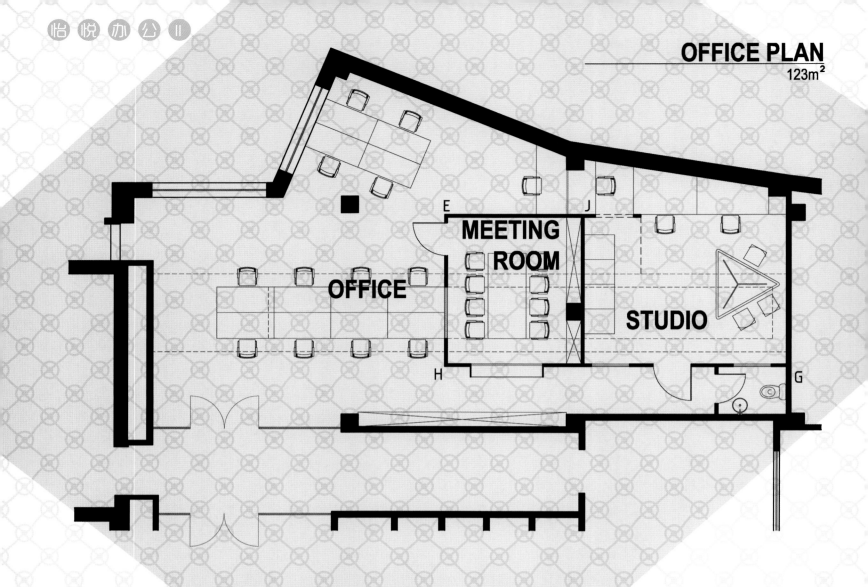

OFFICE PLAN
123m²

MCCANN-ERICKSON RIGA PR-AGENCY

麦肯埃里克森里加公关部办公室

Open AD

设 计 师：Zane Tetere, Rita Saldeniece
建筑面积：125 平方米
项目地点：拉脱维亚里加市

Architects: Zane Tetere, Rita Saldeniece
Area: 125 m²
Location: Riga, Latvia

PR Office is a conceptual continuation for already made McCann Erickson-Riga and Inspired offices, which are located in the same building.
Functional planning is made as stylization of union of inner room and outside—there is formed "courtyard" in which the working places are located, and as seperated block is located "house" with windows and facades, in which is meeting room and a room for media trainings/studio.
The linear composition is followed by overhead lighting under which is planned as working spaces and also the "building", which is functionally inhabitated from all sides there is also integrated windowsill for reading magazines or brainstorming.
Glass/window wall near the distant facade is made transparent for possibility to see all the buildings, not just a part of them.
Considering finishing and furniture, simple materials are adopted, boards, lacquered MDF, grey lacquered blastula, straw color carpet and imitation grass, which are the symbol of stylization of courtyard and meadow.

麦肯埃里克森里加公关部办公室与麦肯里克森里加创意办公室位于同一座楼，在设计理念上延续了后者的创作风格。
此办公空间在功能规划方面遵循空间内外混合原则，即设计师在室内工作区域里设置一个"庭院"，使室内空间就像是分隔开的带有窗户和立面的两个建筑体，建筑体内部分别是会议室和媒体工作室。
整个办公空间的线性组合设计充分考虑了设在工作区和"建筑体"（会议室，媒体室）上方的照明系统，"建筑体"四周均设有功能性飘窗，可以作为办公之余的阅读思考区。
远端立面设有全透明式玻璃窗，以便将建筑景观尽收眼底。出于装修加工与家具的考虑，设计师采用了简单的材料，比如木板、喷漆纤维板、灰色喷漆毛坯、麦秆色仿草地设计的地毯，体现了庭院草地融合的设计风格。

OFFICE PLAN
123 m²

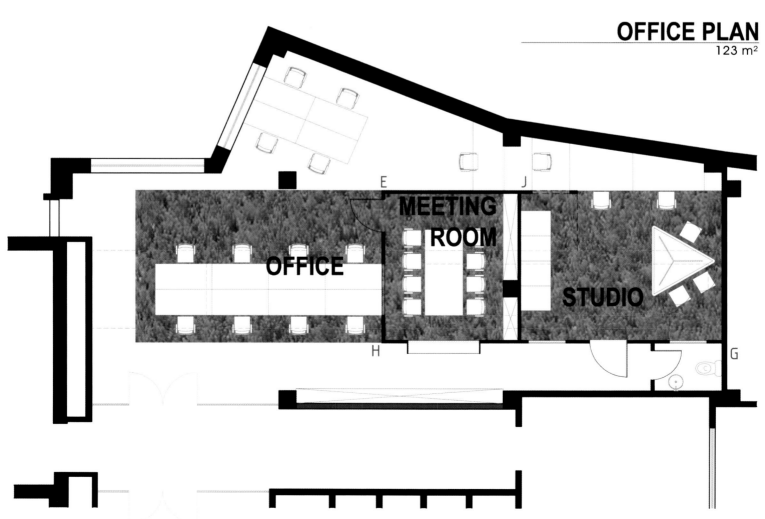

FRAUNHOFER PORTUGAL
弗劳恩霍夫办公室

Pedra Silva Architects

项目团队：Hugo Ramos, Rita Pais, Jette Fyhn, Dina Castro, André Góis Fernandes, Ana Lúcia da Cruz, Ricardo Sousa, Bruno Almeida
建筑面积：1 660 平方米
项目地点：葡萄牙波尔图市
摄　　影：João Morgado

Project Team: Hugo Ramos, Rita Pais, Jette Fyhn, Dina castro, André Góis Fernandes, Ana Lúcia da cruz, Ricardo Sousa, Bruno Almeida
Area: 1,660 m²
Location: Oporto, Portugal
Photography: João Morgado

Our team was selected, through an open competition, to design the new Porto headquarters, located at the Technology University Campus: "Science and Technology Park of the University of Porto (UPTEC)". Our design took into account Fraunhofer's innovative philosophy through a message that is simple, positive and dynamic. Innovative workplace layout and organizational elements from Fraunhofer Office Innovation Center in Stuttgart, Germany, were also an important input to the project, adding another layer to our concept.

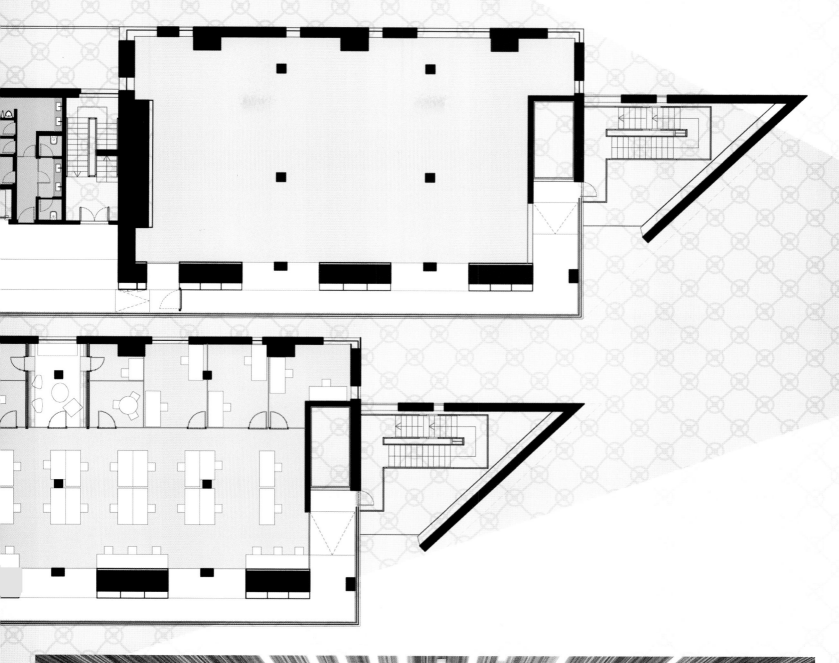

The new research facilities occupy two floors in a new UPTEC building in a total of 1,660 m². Circulation is the project's backbone; all spaces appear along a distribution route located next to the glass facade. This main axis allows access to all different spaces. These spaces, with different functions and sizes, are generated and consolidated through a bold gesture: a waving plane that goes through the open floors, creating different spaces and ambiances. This spatial and visual dynamics are generated by a free plane that travels through the space and by color, which reinforces the perception of different volumes. The waving surface acts, depending on the context, as ceiling, wall or floor of offices and meeting rooms, guaranteeing visual continuity, movement and flow. Another important asset to the project is the introduction of several small social and meeting spaces, named silent rooms, which allow for personal retreat, as well as informal meetings or resting. These spaces are intended to generate a highly creative environment promoting comfort and well being among the researchers.

该设计团队在一次公开竞标中胜出并参与设计位于 Porto 科技大学内的弗劳恩霍夫公司新办公总部。该设计通过简洁，正面且富有活力的手法展示了弗劳恩霍夫的创新哲学理念。公司位于德国斯图加特的分支机构里新颖的办公环境布局以及井然有序的设计元素也被重点应用到总部项目中，为我们的概念设计增添了另外一层高度。

新式的科研设备占据了 UPTEC 新大楼内的两个楼层,总面积达 1 660 平方米。空气流通是这一项目的核心所在;所有的空间沿着玻璃幕墙的侧边分布。建筑核心轴线可连通各个的空间。这些不同风格、结构和尺寸的空间的设计建造灵感都源于大胆的结构尝试,即设计存在于整个开放式楼层的波浪形的平台,营造出不同的空间和氛围。这种空间与视觉上的动感由空间色彩所形成的自由面所产生,增强了不同方位之间的感知力。波浪形外表的呈现需依靠周遭环境,如办公区和会议室的天花板、墙身或者地板,确保视觉上的连续性、动态性以及流畅性。项目的另一个优点是增置了几处小型的社交会面空间,被称为"静室"。这里可作为个人的私密空间,也可用作休息或正式会议。这些空间旨在形成极具创意的环境,为科研人员打造一份舒适安康的氛围。

VIP WING IN MUNICH AIRPORT
慕尼黑机场贵宾休息室

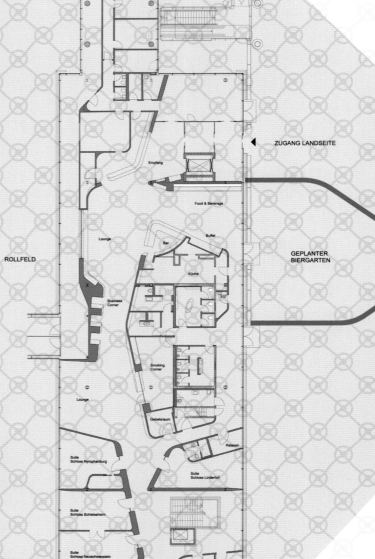

Erich Gassmann, Tina Aßmann

合 作 方：Philipp Hutzler, Andreas Obermüller, Sebastian Filutowski
照　　明：Tropp Lighting Design
建筑面积：1 200 平方米
项目地点：德国慕尼黑市
摄　　影：© Florian Holzherr

Collaborator: Philipp Hutzler, Andreas Obermüller, Sebastian Filutowski
Lighting: Tropp Lighting Design
Area: 1,200 m²
Location: Munich, Germany
Photography: © Florian Holzherr

The VIP lounge is a modern implementation of a special identity, skillfully uniting progress and tradition. It can be experienced with all the senses: haptically—in the use of native woods, typical Bavarian materials such as loden, felt, leather, and broad oak planks; and visually—with maximum use of daylight, selective views of the famous white and blue Bavarian sky, and the proposed beer garden. An island of peace and tranquility in the turbulent working day, a snatch of holiday between appointments, a perfect work oasis equipped with the latest technology, international flair with a Bavarian accent.

In the south wing of Terminal 1, guests have 1,200 m² at their disposal, consisting of four separately usable suites, a central lounge area with gastronomy, business corner, and separate workrooms, rest area, showers and changing facilities, an interdenominational prayer room, as well as a smoking corner. The room conveys a feeling of the alpine mountain world. The falling and rising room recesses determine the conception of the room. The requirements on material and workmanship are manifested at different levels—from the oak-clad doors with their hand-finished brass fittings, the solid wooden tables carved from a tree trunk, to the sensual experience of the fragrant Swiss pine in the relaxation room.

In the reception area, the traveler is greeted by a typically Bavarian motif: A wall of untreated larch shingles covering the rear panel of the reception counter. Comfortable leather sofas and armchairs in the second lounge also invite the traveler to linger awhile. Even from the smoking corner opposite, the large glass window provides a view of the runways. Hand-finished tables of solid ash with finger joints represent Bavarian craftsmanship.

In addition to relaxation at the bar or in the suites, the business corners offer a suitable atmosphere for quiet, subdued work and communication. Alcoves cut into the wall, with felt-cushioned niches and solid-oak tables, are equipped with all the necessary connections for the latest means of communication.

VIP休息室是对特殊身份的一种现代化体现形式，它巧妙地融合了发展与传统的特色。本案设计更可以向贵宾提供全方位的感官体验：触觉上——采用典型的巴伐利亚当地木料、建材，如罗登呢、毡制品、皮革以及宽大的橡树厚木板；视觉上——最大限度地借助日光，精选展现了颇负盛名的巴伐利亚蓝天白云、啤酒花园。如此精雕细琢后，VIP休息室宛如喧嚣工作日中一座难得宁静的"岛屿"、繁忙工作中可贵的短暂休闲假期以及备有最新科技的完美绿洲，充斥着浓厚的国际气息与巴伐利亚风情。

在候机一号楼的南翼，宾客享有1 200平方米的大型娱乐空间。它由四个可分离使用的套间组成，中央的休息区不仅提供美食还设有商务角、单独的工作室、休息区、淋浴更衣室、针对不同教徒的祷告室、吸烟区。房间模拟阿尔卑斯山脉起伏的轮廓，针对功能概念设置壁龛。依据不同层次设置建材和手工艺品——既有橡木里层的大门搭配手工制的黄铜家具、树干雕刻而成的实木桌子，又有带来的芳香感官体验的瑞士松木。

 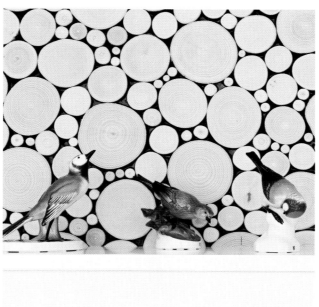 

在接待区，柜台的后方墙身采用未经处理的落叶松砂砾覆盖，洋溢着传统的巴伐利亚氛围来迎接贵宾。第二休息室内舒适的皮革沙发和扶手椅可供游客短暂停歇。即使身处吸烟区的对面，贵宾仍可透过大块的玻璃窗看见跑道。手工制的坚实白蜡木桌子刻有手指关节的图案，在细节处体现着巴伐利亚的手工艺技术。
吧台、套间提供休闲功能，商务角则提供宁静氛围，满足贵宾工作与通讯的需要。
室内配有壁龛和坚实橡木圆桌，均被嵌在墙内，还配备了新型通信所需接口，以备贵宾不时之需。

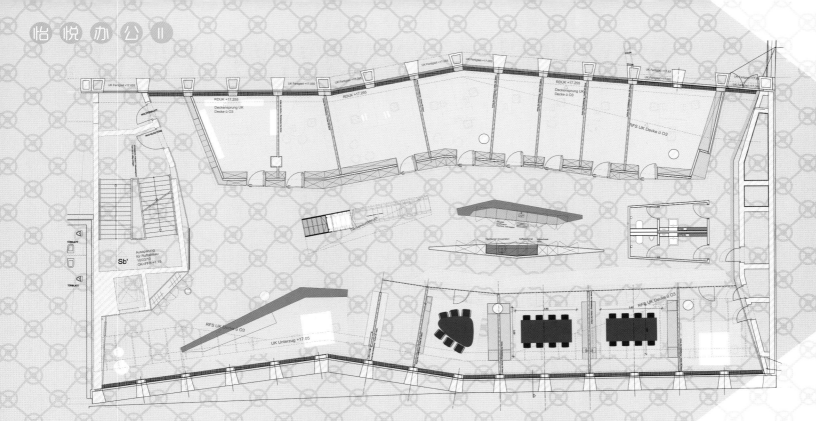

SIGNA HOLDING HEADQUARTERS

思歌娜控股集团总部

Landau+Kindelbacher Architects

建筑机构：Landau+Kindelbacher
建筑面积：1 200 平方米
项目地点：奥地利因斯布鲁克市
摄　　影：Christian Hacker

Architects: Landau+Kindelbacher
Area: 1,200 m²
Location: Innsbruck, Austria
Photography: Christian Hacke

Together with the ceremonious opening of the Tyrol department store, the new headquarters of Signa Holding was opened with a luxurious interior design on Tyrol's top two floors.

In addition to generous office spaces and conference areas, there is also a public lounge where employees can relax and enjoy a fantastic view of Innsbruck. Elegance and understatement, accompanied by an unusual mixture of materials and accentuated lighting, complete this extraordinary building. On the lower floor are the reception, a coffee bar and office areas, while the upper floor is reserved for business activities with conference rooms and the owner's office. This arrangement is reflected in the choice of colors and materials. A sculptural staircase carved out of Tyrolese walnut is a local material reference that optically connects the two function areas. These terminate in a distinctive bar element that encourages communicative exchanges. Terrazzo flooring, precious woods, leather-lined walls and furniture, fine fabrics made of textiles and metals, plus the elaborate workmanship in the various materials, emphasize the clear and fine elegance of the new premises.

随着提洛尔百货商城隆重的开幕，位于顶楼两层新落成的思歌娜控股集团总部，以其豪华的室内面貌迎来了首日的营业。

除了宽敞的办公空间和会议区，还设有公共休息室，那里可供员工放松并享受因斯布鲁克市的绝妙美景。高贵典雅与轻描淡写并存，建材和灯光与众不同地混搭，从而造就了这一卓越非凡的建筑设计。低楼层设有接待处、咖啡吧台和办公区；高楼层则用于商业活动，设有会议室和物业主管的办公室。这样的安排可以从颜色和建材的选用上判断出来。雕刻式楼梯的原料是当地的建材提洛尔核桃木，它从视觉上连接两个功能区域。酒吧则一改之前的风格，这种与众不同的设计元素有助于沟通的需要。水磨石地板、珍贵的木料、皮革衬里的墙身和家具、由纺织品和金属制成的精良织物以及多种材质制成的精细手工品，无不强调出这一新建场所清明透亮的优雅属性。

KUSTERMANN PARK

Kustermann 园区

Oliv Architekten

设 计 师：Thomas Sutor
建筑面积：43 000 平方米
项目地点：德国慕尼黑市

Architects: Thomas Sutor
Area: 43,000 m²
Location: Munich, Germany

This office design shelters passion and creativity. It is a formal and logical work under endeavor that addresses the needs of staff there. The space is a narrative of complex systems which offer beauty and efficiency through tension and decoration.
Office design is required to be logical and user-friendly, suitable to show spatial layout, to take full advantage of spatial resource, to deal with functional needs. Besides beauty, practicability and safety, office decoration more likely intends to stress a kind of aura. The fundamental goal of office design is to provide staff with a comfortable, convenient and healthy space.
In this case, wall color is simple in pure white, which leaves an impression of brightness. The meeting room going with black armchairs is full of formal and rigorous atmosphere, which is good for concentration. A bit of vivid colors like red and orange are used to intersperse the space, it properly gives a little surprise but is not too over. The red glass walls are definitely worth mentioning, they match well with the wooden ceiling and floor, this special and shining combination is quite a characteristic of this office design. Staircases are elegant in dark blue color. The golden reception desk is most eye-catching and perfectly highlights the image and culture of the corporation.

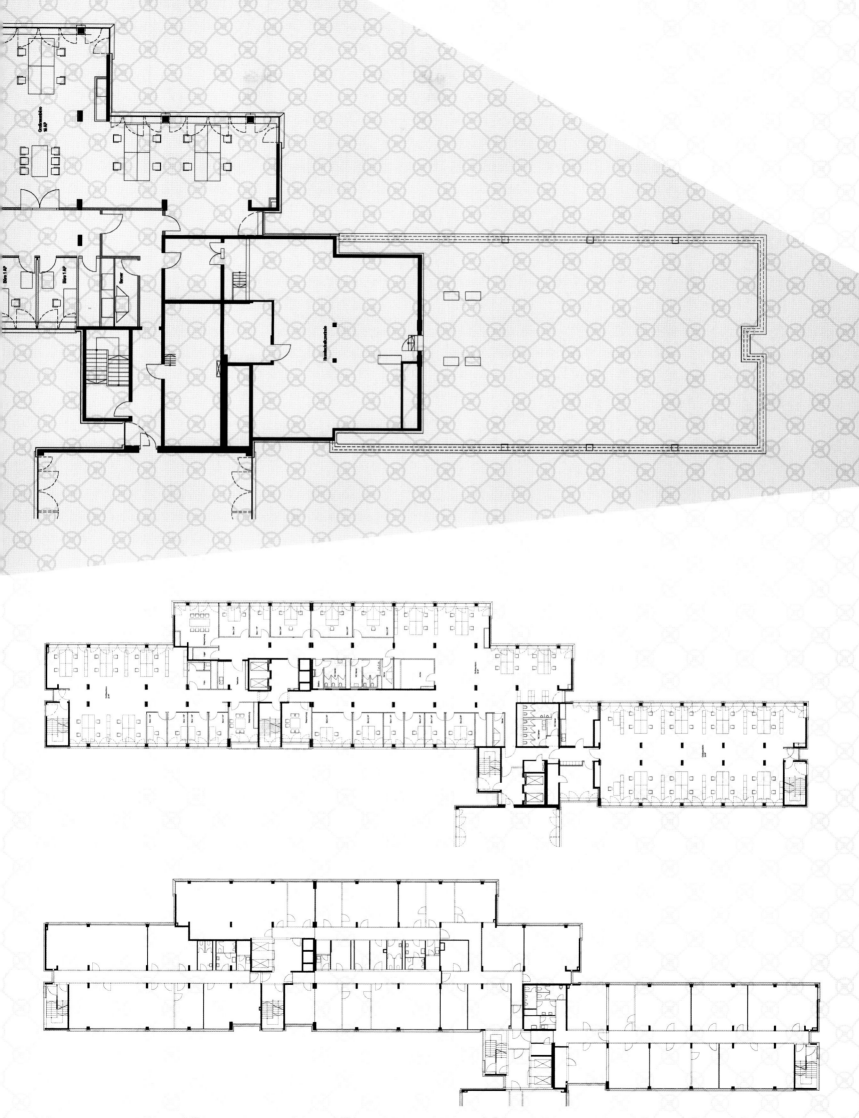

这一办公设计项目承载着设计师的激情与创造力,既有条理又具逻辑性,它凝结着设计师想要满足员工需求的拼搏精神。这一空间体现了标准的综合体建筑,通过空间的张力与装饰呈现出美感与效率感。

办公设计要求合乎逻辑并便于使用,适当地展现空间格局,充分利用空间资源,解决功能上的需求问题。除了考虑美观、实用和安全性以外,办公装饰更注重强调一种氛围。设计的基本目标是为员工提供舒适、便捷且健康的办公空间。

在这一案例中,墙身颜色是简单的纯白色,给人以明亮的视觉感受。会议室配有黑色的扶手椅,充斥着正式严谨的气氛,这更有利于精力的集中。采用少量鲜艳的红、橙颜色点缀空间,恰如其分地营造出惊喜感。红色玻璃墙绝对值得一提,它与木质的天花板和地板相得益彰。这一独特耀眼的搭配成为整个办公设计的特色之一。楼梯采用典雅的深蓝色调。金色的前台最引人注目,完美地突显出企业形象与文化。

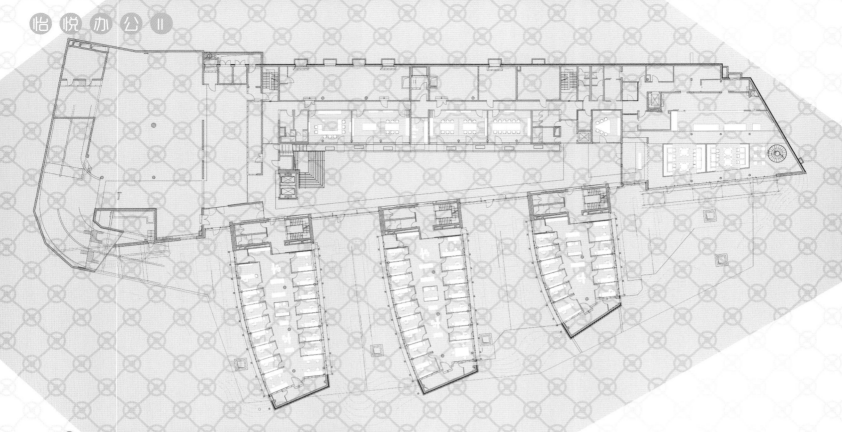

LHI HEADQUARTERS

LHI 总部

Landau + Kindelbacher Architects

总体规划：Mann & Partner
建筑面积：13 000 平方米
项目地点：德国慕尼黑市

General Planning: Mann & Partner
Area: 13,000 m²
Location: Munich, Germany

LHI's new headquarters is located in Pullach, it follows a desire that the campus atmosphere is composed of all facilities, such as conference facilities, casino, office areas, communication areas and green areas, all included in a compact space. Originally using a bulky structure to adapt the building to the environment and spreading over six permeable parts of the building takes place. Tradition, identification and emotion play the important role, and these elements will be implemented in the utilization of local materials and the matching with surrounding landscape.

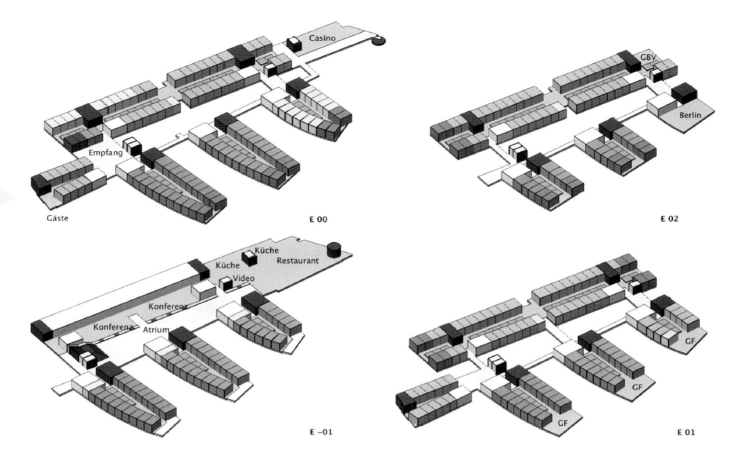

In the center of the building is the spacious entrance hall, it is an event space at the same time. The real "basement" reached by stairs is representative, with a direct access to green areas, and it also can be used as an area for various kinds of conferences. The final part of the hall is the two-story, daylight-filled casino.

The design's goal is to meet the safety requirements of a contemporary work area with a mix of single and combined office areas at all floors. A large central area is a place for both storage and communication. The team zones at the end of the building parts are supplemented by some furniture which welcomes leisure and informal chat.

The selection of materials and colors is cautiously limited. Light natural stone is natural earthy tones and used for the facades and floors. The utilization of oiled oak plus greige and beige fabrics expresses the concept of naturalness. Only the red-color LHI logo is in bright tone.

  LHI新落成的总部位于慕尼黑市南郊——普拉赫镇，它遵循的理念为：园区的氛围是由所有设施共同营造的，因此在这紧凑的空间内涵盖了尽可能完善的设施，如会议设施、娱乐区、办公区、交流区以及绿化区。

  为了顺应建筑周围环境，并且使建筑相互连接的六个部分充分地融合起来，独出心裁地采用大型结构模式。传统、认同感和情感因素在建筑设计中扮演着重要的角色，这些都体现在选用当地建材、融合周边景观上。

  建筑的中央有宽敞的门廊，同时也用做活动场地。真实存在的"地下室"极富特色地由阶梯连接，直通绿化区，也是可供举办各类会议的区域。大堂的最后一处是两层楼高且日光充足的娱乐区。

  设计须满足现代办公区域的安全要求——所有楼层的单间和复合式办公区域须搭配得当。大型的中央区域是可供储藏和交流的场所。建筑尽头的会议区添置了部分家具，可供休闲交流之用。

  建材和颜色的选取则须谨慎有度。天然形成的浅色石块被用于外墙和地板，呈现了自然的大地色调。油质的橡木搭配灰褐色和米黄色的织物，表达出自然主义的理念。唯有红色的LHI标志采用亮色调。

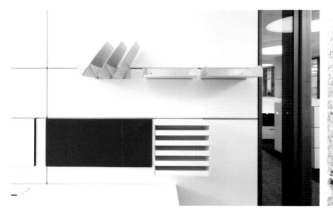 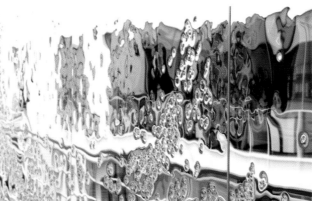 

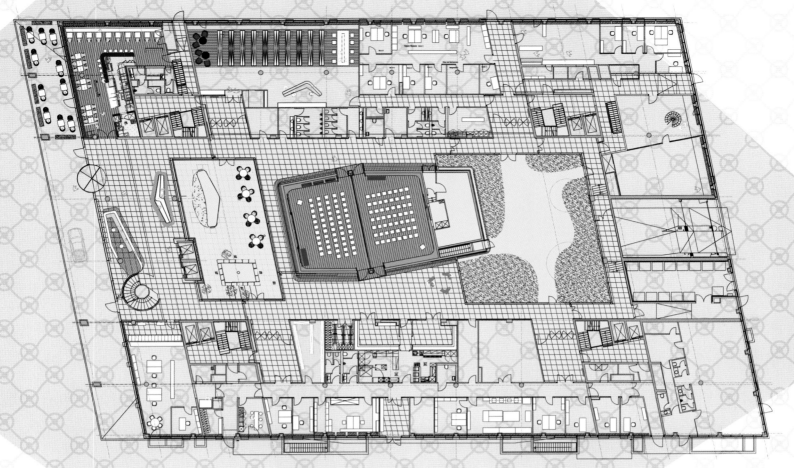

NEW OFFICE DESIGN ICADE PREMIER HOUSE

ICADE 总理府新办公楼

Landau + Kindelbacher Architects

建筑面积：22 500 平方米
项目地点：德国慕尼黑市
摄　　影：Christian Hacker, Werner Huthmacher

Area: 22,500 m²
Location: Munich, Germany
Photography: Christian Hacker, Werner Huthmacher

The implementation of the company philosophy in architectural corporate identity is reflected not only in the entrance lobby, the cafeteria, the conference areas, the auditorium and the management areas, but also in the design of the workplaces. The new style emphasizes perfectly the company's slogan—quality in everything we do. The open character of the building is enhanced by bright, elegant colors in the entrance hall.

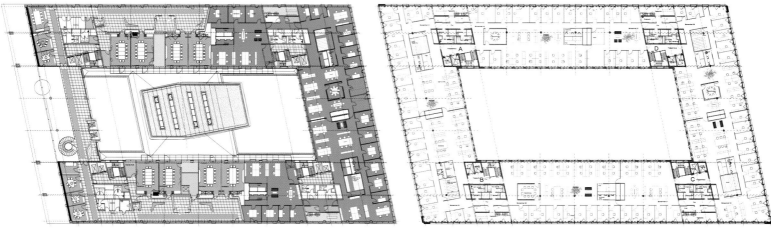

The use of wood, natural stone and Corian as the recurring design elements produces a natural, warm atmosphere. The design concept runs through the entire project; even the elevators are lined with natural stone, and the facades of the entrance hall are adorned with it. As the dominant color, a strong Bordeaux red is chosen for carpets and leather coverings to provide a contrast to the partially illuminated white and the warm wood tones. The auditorium is one of the gems of this project, providing a successful symbiosis between the Corian as the enclosing casing for the interior and exterior and the use of warm wood and technical fabric wall covering for the interior. The library also deserves attention with its division into reception, reading and lending areas. The cafeteria exudes a friendly, open character: both the choice of colors and the furniture invite visitors to linger.

对于建筑企业经营理念的落实与表现不仅反映在前厅、自助餐厅、会议区、礼堂以及行政管理区,还反映在办公区的整体设计上。新风格完美地突显出该企业的品牌标语——保证品质,尽善尽美。前厅那明亮优雅的颜色烘托并提升了建筑的开放式特色。
采用木料、天然石块和可丽耐(美国杜邦公司制造的人造大理石)材质这些再生设计元素以营造一种自然温暖的氛围。设计理念在整个项目设计中得以体现;即使是电梯的两旁也衬有天然石块,令前厅的外墙增色不少。浓艳的波尔多酒红色作为主色调被用于地毯和真皮椅套,与局部的明亮白色和暖木色调形成鲜明对比。礼堂是这个项目的亮点之一,它成功地令这二者和谐并存,即用于室内外的可丽耐材质保护套以及用于室内由暖木和精致织品制成的墙身保护套。图书馆同样引人注目,它分为几个部分:接待区、阅读区和借阅区。自助餐厅洋溢着友好、开放的气息,精心挑选的颜色和家具传达着"欢迎再来"的信息。

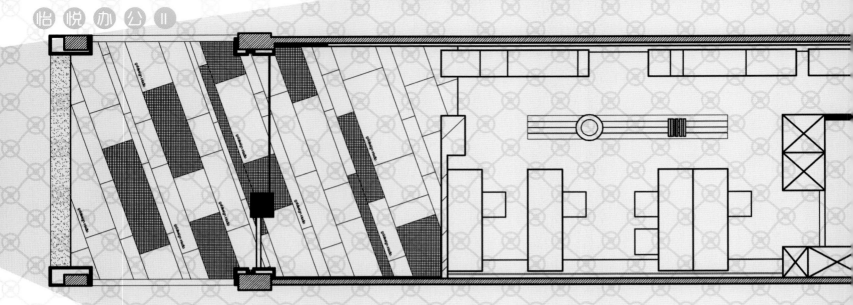

INTENTION · PRIMARY COLORS
初·原色

Goldesign Studio

设 计 师：凌志谟
建筑面积：200 平方米
项目地点：中国台湾省桃园市
主要材料：大理石，不锈钢板，镀钛板，黑铁烤漆，黑铁板镭射，
　　　　　配电管，玻璃，橡木
摄　　影：吴启民

Architect: Zhimo Ling
Area: 200 m²
Location: Taoyuan, Taiwan, China
Main Materials: Marble Stone, Stainless Steel Plate, Titanium Plate, Dark Iron Baking Varnish,
　　　　　　　 Dark Iron Laser, Electrical Pipe, Glass, Oak Wood
Photography: Qimin Wu

"Creative Origin" is the first idea coming from the whole space, hoping creativity "stay hungry". As the saying goes, "stay hungry, stay foolish", we hope to have bigger breakthroughs and make greater impacts on designers, not just aiming for style or other form languages.
The style of aesthetics speaks for itself, and high-profile showing off is not necessary. Keeping designers' passions and original intentions, through the space in black and white, bringing designers back to the beginning, tranquilizing their minds, and getting rid of contemporary color and catch phrases will stimulate more possibilities for creativity.

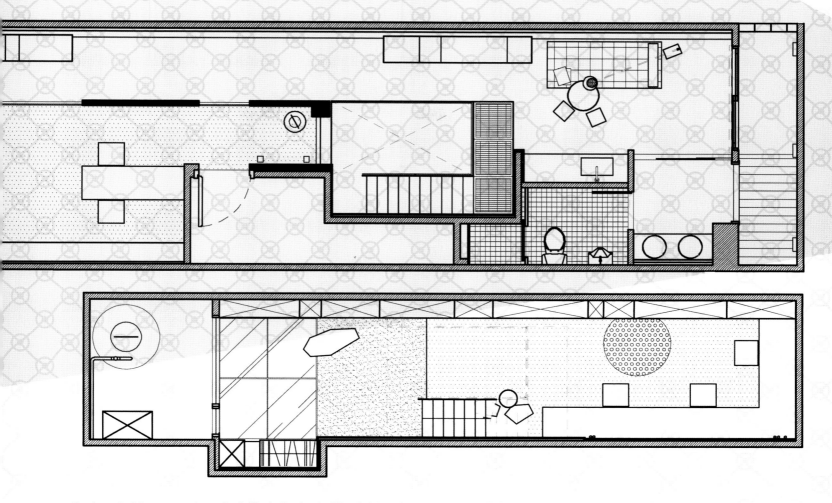

The layout of the space shows flexibility. In the front of the first floor, the designer's workspace is suitable for both individual and team work. Designers can fully make use of this area. For example, the sliding office tables can be combined to become a big conference table for group meeting. Besides, there are also sliding steel shelves which create a more colorful background.

goldesign studio

The wall was designed for bringing about humanistic images to the space. The reading gallery composed of sliding bookshelves makes knowledge and inspirations move forward with the route. Walking is no longer moving your body because there are places for reading and imagining everywhere. This is also the main point in spatial design: you do not have to wait for inspirations as every space can trigger your ideas of designing. In addition, open-riser marble stairs reduce the sense of alienation and allow drawing area downstairs to interact with free discussion area, flexible reading space, and romantic workspace upstairs. It is the main point of this design.

❝ "创意原色"是从整体空间中引申出的第一理念，即希望创意能一直停留在"饥饿的状态"。常言道"求知若饥，虚心若愚"，设计才能有更大的突破，带给设计工作者更大的思维冲击，而不是一味追求式样或其他的语言形式。
美学形式本身不言而喻，高调并不是唯一的表现形式。保持设计者对设计的热情初衷，采用低调的中性色彩让设计者回归原始、沉淀自我，借此摆脱潮流色彩与流行语汇，才能激荡出更丰富的原创构思。
空间的分割规划极具灵活性，一楼靠前位置的设计工作区可供设计师独立或群聚工作。设计师可以充分利用这一区域，比如滑动的办公桌可以组合成巨大的会议桌供团队开会使用，带有滑轮的书架可使空间背景变得丰富多彩。
墙面设计给空间添上些许人文情怀。活动书架构成长廊，知识、灵感伴你前行，身处在满是阅读和创意的想象空间内，走路已不再是单纯的身体移动。这种设计强调了设计师不必再等待着灵感的出现，因为处处都是可以激发创意的想象空间。此外，透空的大理石楼梯，减少了空间隔绝感，使楼下的绘图区也能和自由讨论区有所互动，营造了灵活的阅读空间和浪漫的工作气氛，而这种轻松的环境正是本次设计的重点。

MPD OFFICE

MPD 办公室

StudioLAB

设计团队：Matthew Miller, Ryan Ho
建筑面积：280 平方米
项目地点：美国纽约市
摄　　影：In-house

Design Team: Matthew Miller, Ryan Ho
Area: 280 m²
Location: New York, USA
Photography: In-house

The 280 m² space was designed for a boutique creative agency located in the Meat Packing District of NYC. Existing cubicle walls that cut up the space and shelter natural light were demolished to give way to an open plan with a loft like feel. Simple, frameless glass volume enclosures were used for staff offices to allow daylight to permeate through the space along with the white washed floor to maximize brightness. Custom stainless steel framed desks were built along with a custom designed high gloss lacquered reception desk. The clean-lined forms and modern material palette reinforce the company's minimalist aesthetic approach.

Materials and products used include: Starphire frameless glass partitions and sliding doors using CRL Lawrence door hardware; custom-designed high-gloss white lacquer reception desk with integrated light box; custom finished white oak flooring with white-wash stain; staff workstations with stainless steel framed bases and custom ApplePly tops with exposed edges; office kitchenette using Ikea high gloss white lacquer cabinets; bleached white oak millwork built-ins; cooper lighting architectural indirect suspension lamps; SheerWeave light diffusing mechanical solar shades; stripped and exposed steel doors with clear lacquer finish, including elevator door.

这一280平方米的空间是为一家创意精品店设计的，它位于纽约市的肉食品工业区。原本的隔断不但分隔了空间，而且遮挡了自然光，现将其拆除，为富有阁楼感觉的开放式规划保留空间。员工的办公区采用简单无框架的玻璃围墙，日光可以洒满整个空间，在白色可拖洗地板的衬托下最大程度地亮化室内空间。定制的不锈钢镜框桌子与定制的高光漆前台相互辉映简洁的线条形式和现代的建材色调强调了企业极简抽象派的审美取向。

设计采用的建筑材料包括：Starphire 无框架玻璃隔板和装有 CRL 劳伦斯硬件的滑动拉门；配有完整灯箱的特制高光白漆前台；带有石灰斑点的特制白色橡木地板；员工工作站内的不锈钢镜框基底和边缘外露的特制 ApplePly 顶盖；办公区内小厨房采用的宜家高光白漆收纳柜；雪白的橡木制品内置插件；库伯的建筑悬浮灯；可漫射太阳光的薄纱织物灯；包括电梯门在内，条状、外露的漆涂料铁门。

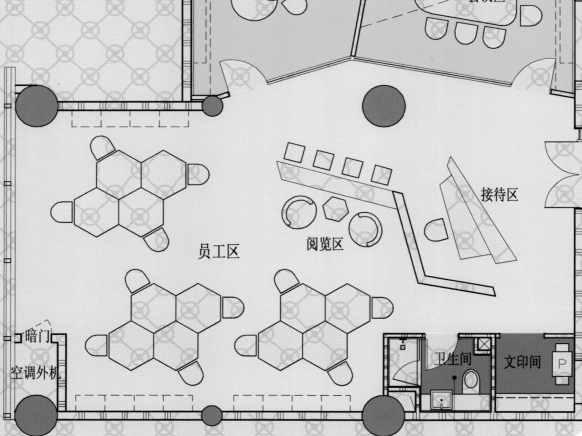

CHONGQING CHUANGHUI FIRST OFFICE MODEL HOUSE

重庆创汇首座大厦办公样板房

Zen Chi Design

设 计 师：王哲敏
设计团队：王逸飞，严宏伟，周林虎
建筑面积：170 平方米
项目地点：中国重庆市
摄　　影：王哲敏

Architect: Wilson Wang
Design Team: Lucifer Wang, Andrey Yan, Tiger Zhou
Area: 170 m²
Location: Chongqing, China
Photography: Wilson Wang

Many creative studios regard office space in the form of LOFT as the prior choice while considering the office location. Then the Chongqing Chuanghui First Office Model House invested by the Ruian China Hui entrusts the interior designer Zhemin Wang (Wilson) of extremely creativity. Zhemin Wang breaks through the conventions. He sprinkles the creativity to his heart's content whilst making full use of the floor height advantage to built this set of a household type into a creative photography studio, which is inseparable with the fact that Zhemin Wang himself is a professional photography enthusiast. The overall space is mainly based on jumping passionate rosy red and the clean, smooth white color tone, keeping the original cement color, which has a strong sense of industry and is also fashionable. He makes use of the grid and the glass to separate different functional areas such as photography area, office lobby and the executive level office area. In the executive level office, Zhemin Wang chooses largely soft decorations which have his own individual labels "metro classic" style to decorate the space such as the leathery sofa and rudder pedal, exaggerated chandelier, and monogrammed carpets, etc. The model house not only successfully promotes the sales of the property office but also attracts a large number of advertising companies.

LOFT 形式的办公空间是许多创意工作室首选的办公地点，瑞安中华汇投资建设了重庆创汇的首座办公样板间，该项目委托极富创意的室内设计师王哲敏（Wilson）设计。王哲敏敢于创新，在尽情挥洒创意的同时，又充分利用了该空间的层高优势，将这套 A 户型建筑打造成一间创意摄影工作室。当然，这和王哲敏本身具有的专业摄影水准是分不开的。整体空间选用奔放热情的玫红色和干净利落的白色为主基调，地面保持原有的水泥色，工业感十足又不失时尚感。利用网格和玻璃作为隔断，区分开不同的功能区域，如摄影区域、办公大堂以及领导层办公区等。在领导层的办公室里，大量采用富含个人"都汇经典"风格的软装饰来点缀空间，如皮质的沙发和脚蹬、夸张的吊灯以及字母组合的地毯等等。该样板房不仅成功地增加了该产权式办公室的销售额，更吸引了大量广告商前来洽谈。

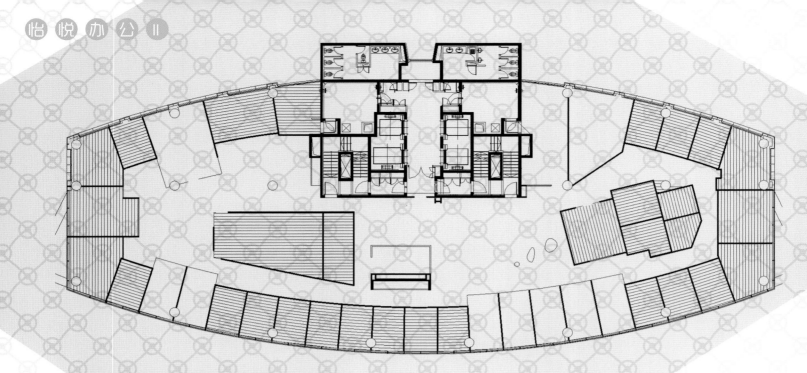

YBRANT DIGITAL

Ybrant Digital 公司

StudioLAB

建 筑 师：Moran Ben Hur, Industrial
设 计 师：Omri Amoyal
项目团队：StudioLAB
建筑面积：1 000 平方米
项目地点：以色列荷兹利亚市
摄　　影：StudioLAB

Architects: Moran Ben Hur, Industrial
Designer: Omri Amoyal
Project Team: StudioLAB
Area: 1,000 m²
Location: Herzeliya, Israel
Photography: StudioLAB

Ybrant is a marketing digital media worldwide office, with a young attitude located in Herzeliya. The office is a 1,000 m², 7-level flat with an ocean view.
The main concept of the design was to create an office with a lively center that is full of light, color and ocean view. We planned the place with inner open spaces that were used for the entrance, circulation and waiting area. Around the center, we placed the working area that was defined as the light and view source for the center.
At the entrance is a monumental front desk that is made out of chopped wood planks. In the waiting area we placed a white cement table and sitting rock that were custom made for this office.

Ybrant 位于以色列荷兹利亚市,是一家年轻态国际化的数码媒体办公机构。办公室地处1 000平方米、7层高的海景公寓内。
设计的主旨理念是创建一个充满活力,五光十色且附有海景观赏区的中心区域。设计师将这个内部的开放式空间用作入口、通风区和等候区。办公区围绕着中心区,并为中心区提供了光源与景观资源。
入口处设有大型前台,由支离交错的厚木板制成。在等候区放置了为办公室特制的白色水泥桌和平置坐石。

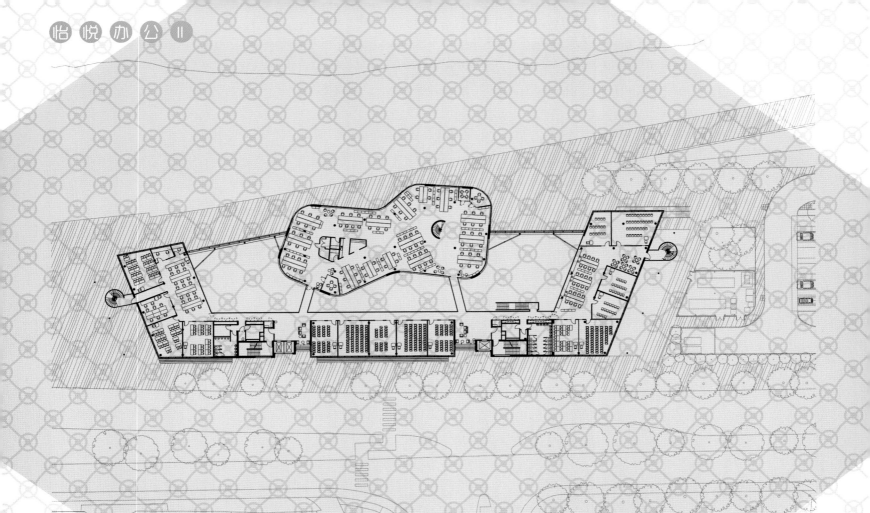

NEWPORT CITY CAMPUS, UNIVERSITY OF WALES

威尔士大学新港城校区

BDP Architects

建筑机构：BDP 建筑事务所
承 包 商：Willmott Dixon
建筑面积：12 000 平方米
项目地点：威尔士格温特郡纽波特市
摄　　影：Martine Hamilton Knight

Architects: BDP Architects
Contractor: Willmott Dixon
Area: 12,000 m²
Location: Newport, Gwent, Wales
Photography: Martine Hamilton Knight

The new campus for University of Wales, Newport, is set to play a key role in transforming Newport's riverside and regenerating the city as a whole. Recently completed on the waterfront in the heart of the city, the building is designed to be visually transparent to showcase the work of the university and to foster outreach.
The campus features an innovative hothouse at its core where artists, business researchers and entrepreneurs can work together and spark off each other to create new commercial ideas. Accessed by walkways, it is raised within the heart of the central space and floats above the tiered resource and refectory space. The organic shape of the hothouse creates a sense of territory, and offers fantastic views and daylight for all users. The ground level podium houses preview theaters, lecture spaces and a studio performance space which can open onto a riverside piazza.
Sustainability was a key consideration underpinning the whole design, and a condition of the funding from the Welsh Assembly Government was that the campus building should achieve a BREEAM Excellent rating. The new campus has achieved a 14.5% improvement on Part L and BREEAM Excellent without heavy investment in low or zero carbon technologies. BDP was appointed architect, landscape architect, interior designer and graphic designer for the £26m project.

位于纽波特市的威尔士大学新校区在纽波特河畔的改造及其城市的整体化进程中起着决定性的作用。近期于市中心海滨区域竣工的这座建筑，视觉上非常通透，直接展示大学取得的成果，促进其更大范围的发展。
校园中心的温室极具创新特色，艺术家、商务研发人员以及企业家在这里共同奋斗，互相激励去创造全新的商业理念。它地处中央空间突起的核心地带，设在分级资源库和食堂的上方，可步行进入。校园的不定型轮廓营造出一种区域范围感，为人们提供极佳的观景角度和日照资源。一楼设有预映影院，演讲空间和可直通到河畔广场的表演工作室。

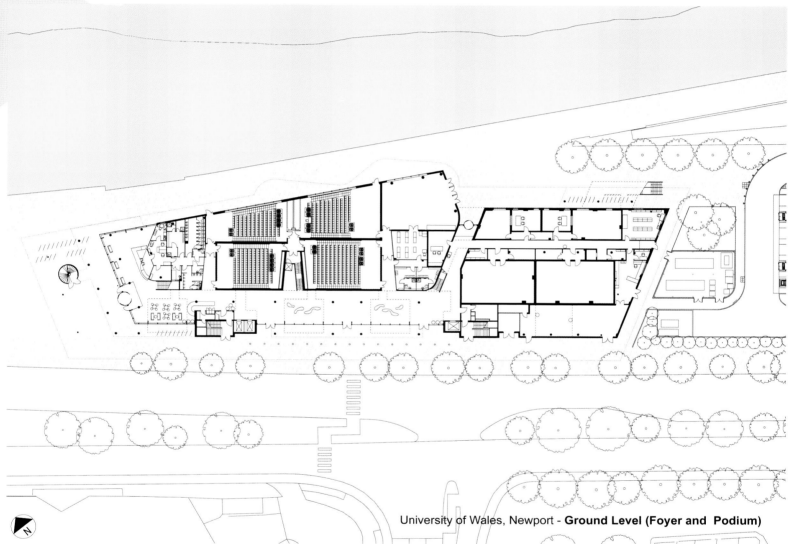

University of Wales, Newport - **Ground Level (Foyer and Podium)**

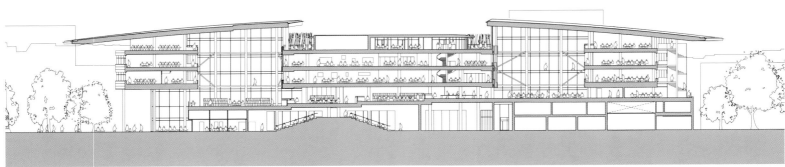

可持续性设计是支撑整个设计的关键,若想获得威尔士议会政府的资金支持,此校园的建设必须依据英国建筑研究院环境评估方法获得优秀等级。新校区将评估方法中 L 部分的数据标准提高了 14.5%,且获得了期望中的优秀等级,省去了在低碳或零碳技术方面的大笔投资经费。BDP 建筑事务所被指定全权负责建设、景观规划、室内设计以及平面设计,整个项目耗资 2 600 万英镑。

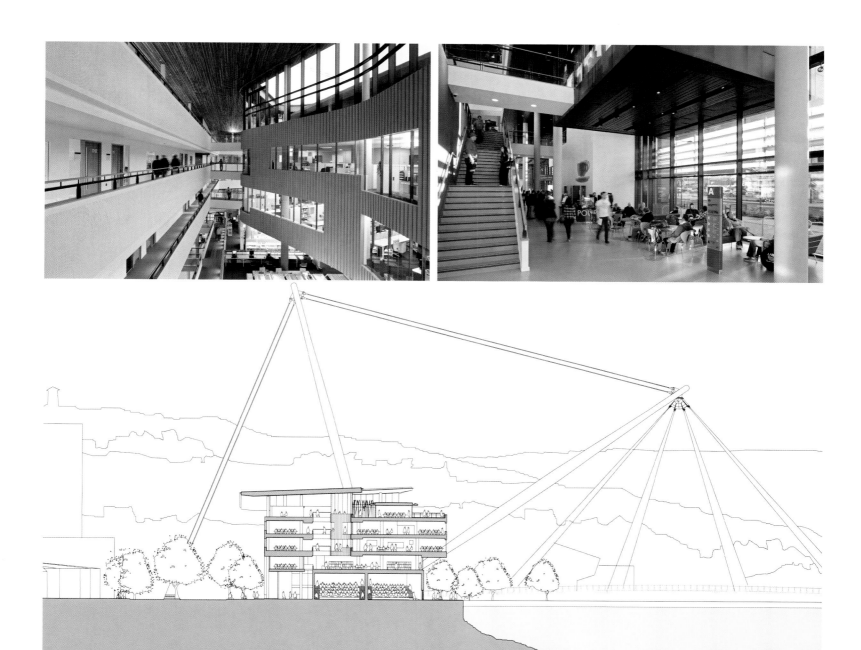

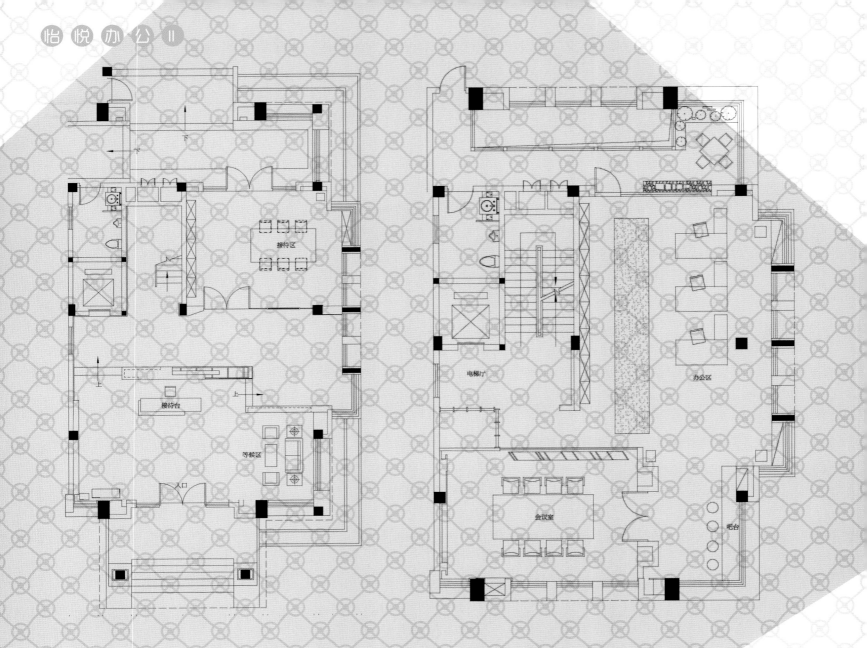

ZHONGQI GREEN HEADQUARTERS · GUANGFO BASE OFFICE

中企绿色总部·广佛基地办公室

Guangzhou C&C Design Co., Ltd.

建筑机构：C&C 设计有限公司
设 计 师：史鸿伟，彭征
建筑面积：800 平方米
项目地点：中国广东省佛山市
主要材料：大理石，复合实木地板，黑色镜面不锈钢

Architects: C&C Design Co., Ltd.
Designers: Hongwei Shi, Zheng Peng
Area: 800 m²
Location: Foshan, Guangdong, China
Main Materials: Marble, Composite Real Wood Floor, Black Mirror Stainless Steel

Zhongqi Green Headquarters · Guangfo Base Office is located in the core district Guangfo—east of the Lishui Town, South China Sea District in Foshan. The total area is 300,000 m² and the construction area is more than 500,000 m². The project consists of ecotype independent office, LOFT office, apartment, 5-star hotel, business club, leisure shopping street.

中企绿色总部·广佛基地位于广佛核心区域——佛山市南海区里水镇东部，总占地面积 30 万平方米，建筑面积 50 多万平方米。项目由生态型独栋写字楼，LOFT 办公，公寓，五星级酒店，商务会所，休闲商业街组成。

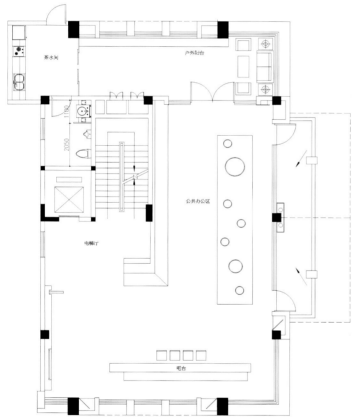
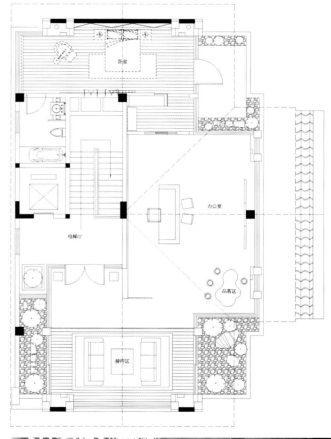

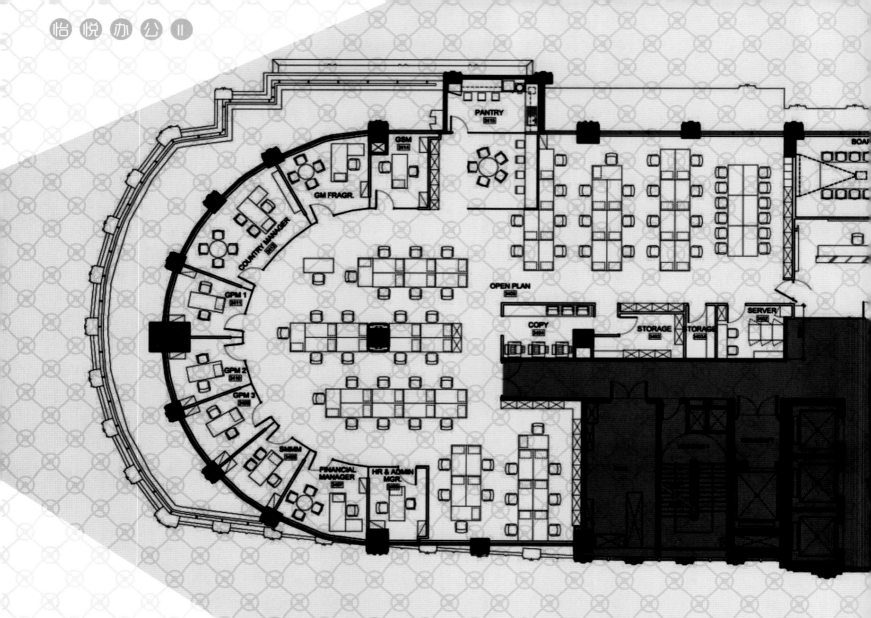

PARFUMS & BEAUTÉ

PARFUMS & BEAUTÉ 公司

dwp

建筑面积：900 平方米
项目地点：泰国曼谷市

Area: 900 m²
Location: Bangkok, Thailand

World-class architecture and interior design firm dwp designed and built the new offices of Parfums & Beauté, the largest retailer of designer on fragrances and beauty products in Thailand. Relocating 70 staff at the enviable top floor of Athenée Tower in Bangkok, the space had to cater to high-end clientele and be a blank canvas, to best feature every brand details. Luxury, sophistication and flexibility were key, as changing feature brands was not to clash with the interiors, which had to be convertible. The client spaces, such as the reception, boardroom and training rooms, had to allow for product launches, inviting high-profile guests and hosting press.

The desired effect was achieved by creating a luxurious palette of high-gloss white in stone and tile, against rich chocolate brown ceilings and dark wood veneers. A reception desk in white stone encased in a lighted glass box offered a hotel concierge like feel. Adding classic Barcelona chairs in white heightened the simple, yet elegant look. Acute attention to detail saw a stunning array of retail displays, as seemingly simple glass boxes that jut out the wall. Yet on closer inspection, the backlit glass boxes had white magnetic walls for flexibility in replacing magnetic logos for the multiple, changing visual merchandising of key featured brands.

The office area received a playful use of the dark wood with a splash of green, within office doors, desk screens, vinyl stickers and paint on walls. A light cheesecake yellow paint contrasts against the patterned grey floor, while thoughtful wallpaper-like pattern paint was achieved in the executive offices, using a single tone of matte and gloss mink-color paint. The lavish boutique hotel ambience does justice to the brands and the company, generating a sense of pride for staff.

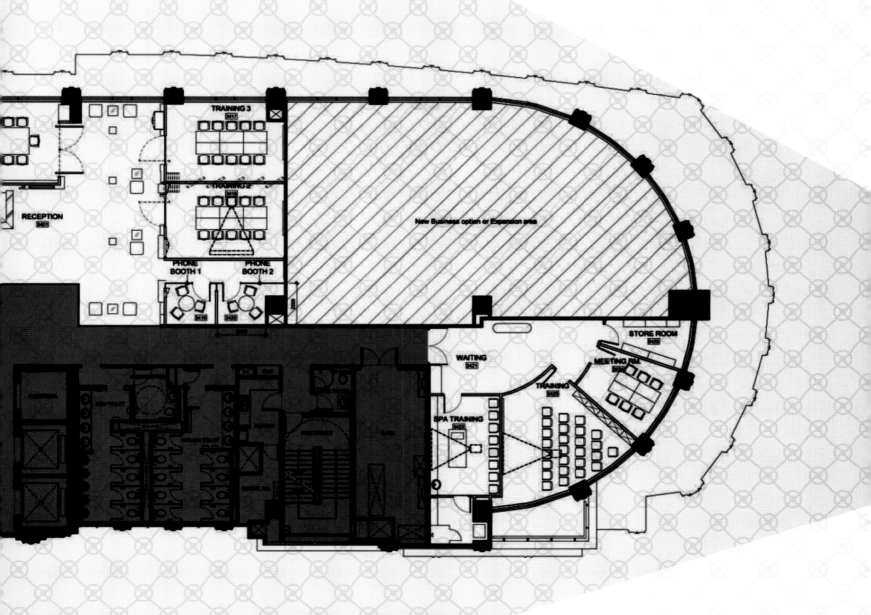

　　世界级的建筑与室内设计机构 dwp 为 Parfums & Beauté 设计并打造了全新的办公场地。Parfums & Beauté 是泰国最大的香水和美容产品零售商。值得称羡的是，70 名员工迁至泰国雅典塔的最高层。空间需要迎合高端的顾客，带给人一种白色帆布的视觉效果，从而最佳地突显出每个品牌细节。奢华、温文尔雅与灵活多变是关键，要知道改变特色品牌与其内涵并不矛盾，而是一种同等意义的更替。客户空间譬如前台、会议室和培训室，均为产品的发布提供优越的保障，以此吸引高端的客人，也便于举办新闻发布会。

华丽高光白色颜料修饰的石料和瓷砖，与巧克力棕色的天花板和深色的木质箱面板形成对比。白石材质的前台以清透的玻璃箱层围裹，营造出类似酒店大堂的感觉。款式经典的白色巴塞罗那座椅兼具提亮的效果，使整体显得既简单又高贵。对细节的细腻处理成就了极富魅力的店面视觉效果。凸出墙身的简洁玻璃箱就是一例，走近观察，背光的玻璃箱有着白色的磁壁，在不同品牌进行特色视觉营销时，可灵活地更换多种磁性商标。

办公区趣味性地使用了深色木料搭配少许绿色作为点缀，在办公大门、办公桌屏风以及墙壁涂料和乙烯基贴纸上均有迹可循。如奶酪蛋糕般淡黄色的墙漆与带图案的灰色地板形成鲜明的对比。行政办公室经过反复规划之后选用了壁纸型涂漆，涂漆采用单色调冰铜和光泽奶白涂料。贵气精致的酒店氛围彰显了品牌和公司端正高档的形象，使员工充满自豪感。

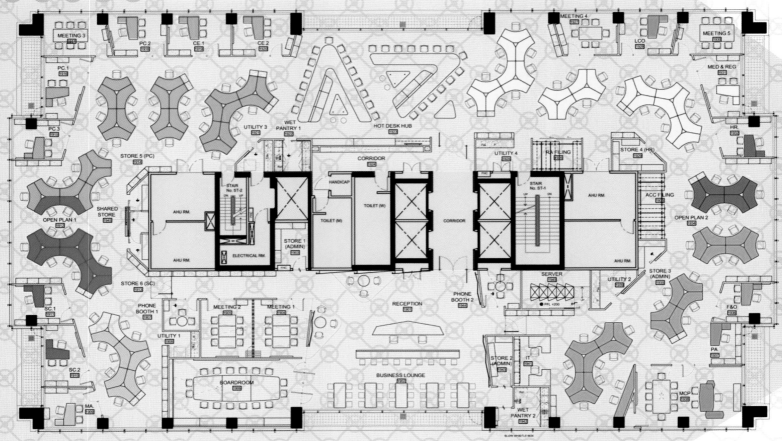

ASTRA ZENECA HEADQUARTERS

ASTRA ZENECA 总部大厦

dwp

建筑机构：dwp
建筑面积：1 400 平方米
项目地点：泰国曼谷市

Architects: dwp
Area: 1,400 m²
Location: Bangkok, Thailand

World-class architecture and interior design firm dwp created the concept for the relocated corporate head offices of Astra Zeneca Thailand, in Bangkok's CBD. Having researched the new offices of Astra Zeneca worldwide, dwp understood the key corporate principle of one of the world leaders in pharmaceutical medicines, equipment and treatment: "Health Connecting Us All". Astra Zeneca was very keen on healthy surroundings, to foster better communication, connectivity, creativity and a sense of community and pride.

Strong branding was essential, using the key Astra Zeneca colors of purple, yellow and violet, to set against a white organic backdrop. A sense of community was achieved by providing a large breakout space, the business center for visitors and staff, in the heart of the office at the reception hall, with the boardroom just adjacent, opening up with large, flexible swivel doors. A violet angular reception desk obtained an angled glass back wall, allowing in natural light. Modern seating in white metal, with yellow upholstery set on carpet inserts, in matching violet. Creating a muted tone of ceramic tile in light and dark wood pattern offered easy maintenance. In the open-plan office space, a specially-designed hotdesk hub of two large counters, shaped A&Z, was reflective of the company's logo. Integrated in this hub are stations for 24 staff and two large sofas, as well as coffee tables and stools. The cellular offices received angular walls for a sense of modern dynamism. Angular aluminum door frames added whimsy and 120° white desks gave an organic feel, while yellow pedestals mimicked the brand, complimented by custom graphic walls.

国际建筑室内设计机构dwp为泰国企业Astra Zeneca起草设计方案，计划在曼谷商业中心区为其重置办公总部。Astra Zeneca作为全球医药、医学器械、医学疗法领域的佼佼者，始终坚持"健康与我们息息相关"的理念。在研究分析Astra Zeneca众多国际分公司后，dwp充分理解了企业的核心守则。Astra Zeneca十分重视健康有序的工作环境，以便更好地交流、接洽和创作，培养团队感和自豪感。其计划实质在于扩大品牌认知度，因此Astra Zeneca的主色采用紫色、黄色和紫罗兰色以平衡白色的主背景。在办公区中央的接待大厅设置大型办公隔间，作为商务中心供访客和员工使用，提升团队感。商务中心与会议室相连，设置灵活旋转大门，对外开放。带有棱角的紫罗兰色前台设有梯形角度的玻璃后墙，方便自然光照射入内。白色金属材质的现代化座椅，配以嵌入地毯的黄色装饰品，与紫罗兰色前台相得益彰。瓷砖的木质图样深浅不一，营造出柔和的氛围，并方便维修保养。开放式办公空间内摆放一款精心设计的A&Z造型摆设品，置于两个大柜台的中心，象征着该企业的商标。可容纳24人的办公区，2套大沙发，咖啡桌椅穿插于此。分格式办公室墙身装饰棱角，为空间增添了一份现代活力。带棱角的铝制门框颇具趣味感，120°仰角的白色桌子给人以整体感。而黄色基座仿制名牌品，为专门定制的图案墙增添了色彩。

UAWITHYA CORPORATE HEADQUARTERS

Uawithya 公司总部大楼

dwp

建筑面积：750 平方米
项目地点：泰国曼谷市

Area: 750 m²
Location: Bangkok, Thailand

Uawithya is a leading name in the quarry and equipment business in Thailand. They recently commissioned world-class architecture and interior design firm DWP to design and build their corporate headquarters on Wireless Road, in central Bangkok.

As an internationally respected company, the office was specifically designed to showcase the numerous products and services the company provides to its clients, as well as provides an efficient, comfortable working environment for staff. A bold statement in the design was conceptualized by using the corporate color red, as the main accent point. This was then set it juxtapose with a strong masculine palette of steel and stone, to convey the nature of the business. There is an evident air of transparency and openness throughout. Additionally, bold graphics were utilized to create an inspirational buzz about the business within the space, and highlight the pride of the company in its past, present and future.

  Uawithya 是泰国有名的经营采石场设备的企业。他们近期委托世界顶级建筑和室内设计公司 dwp 为其设计建造他们位于曼谷中心地带的公司总部。
  作为一个享有国际声誉的公司，Uawithya 的设计既可以向客户展示其丰富的产品和完善的服务项目，又可以提供员工一个舒适高效的办公环境。此案中一处大胆的概念化设计是大面积采用红色作为基准色调，这样与公司主营的钢石业务的阳刚强硬的形象相互作用，为其融入了一些自然柔和的色彩。整个空间会有一种透明开放通畅的感觉。此外，夸张大胆的图形应用让人仿佛感受到了公司运作中激动人心的轰鸣，也突出强调了企业那引以为傲的过去、现在和将来。

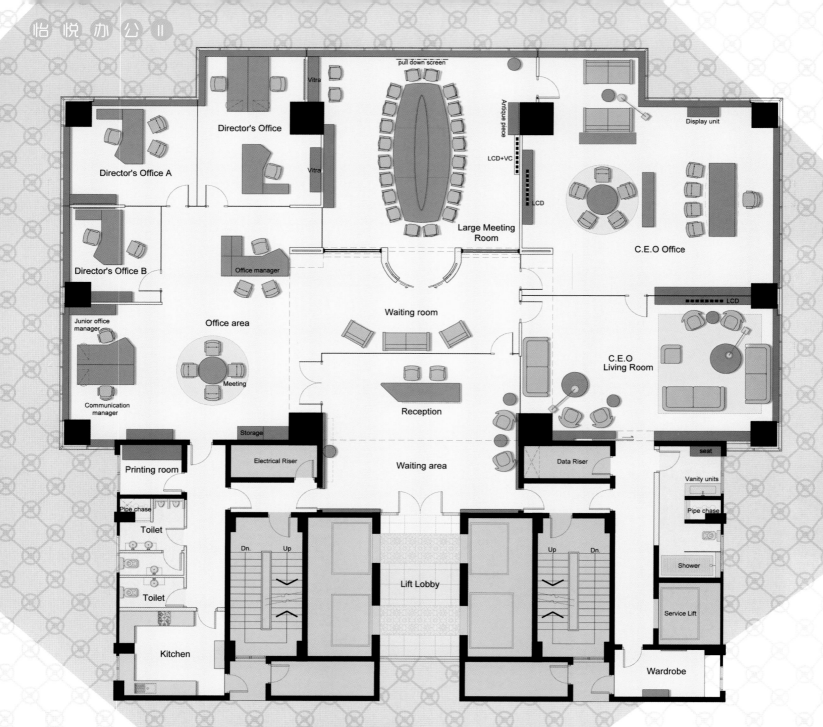

ZAIN HEADQUARTERS

Zain 总部大楼

dwp

建筑面积：11 000 平方米
项目地点：巴林泽夫市

Area: 11,000 m²
Location: Seef, Bahrain

Zain is a telecommunication company in 24 countries in the Middle East and Africa, employing 15,000 staff and having some 64.7 million customers. For the relocation of their group headquarters from Kuwait to Bahrain, Zain engaged world-class architecture and interior design firm dwp, to shape the interiors of a new 11,000 m², 22-storey building in Seef. To accommodate 650 staff, the design by dwp had to respect concept guidelines by Future Brand.
The work involved an upgrade of all existing building services and finishes: lift lobbies, lifts, kitchens and toilets, in addition to offices and client areas. Each floor received its own graphics, relating to its function: finance, IT etc. These graphics were designed by Future Brand, for use in Zain corporate offices. The furniture was primarily from Vitra, with classic pieces from the Eames collection, and workstations were by Hayworth.
Having won the Zain Headquarters project in a competitive bid, dwp were subsequently engaged to provide service for several more Zain projects, including 3 in Africa.

Zain 电信公司在中东和非洲 24 个国家均有分部，雇用员工达 15 000 名，现有 6 470 万名客户。由于集团总部从科威特搬迁到巴林，Zain 公司邀请了世界一流建筑室内设计公司 dwp，为其位于泽夫市的 22 层高、11 000 平方米的新办公楼打造全新的办公环境。除了要求能够容纳 650 名员工，dwp 的设计必须与 Future Brand 企业品牌规划公司所提供的企业形象相协调。

这项任务需要优化所有现有设施与装修效果，除了包括办公空间与客服中心的改造，还包括电梯大堂、电梯、厨房和卫生间的优化。每层楼都具有与其功能相搭

配的独特平面图形，如财务部、资讯科技部等，这些平面图形由新未来品牌公司为 Zain 办公室设计。家具主要来自 Vitra，是 Eames collection 的经典之作，工作台则是 Hayworth 的。

dwp 赢得了 Zain 总部大楼的设计竞标之后，相继参与了 Zain 一些其他项目，其中就包括非洲的 3 个项目。

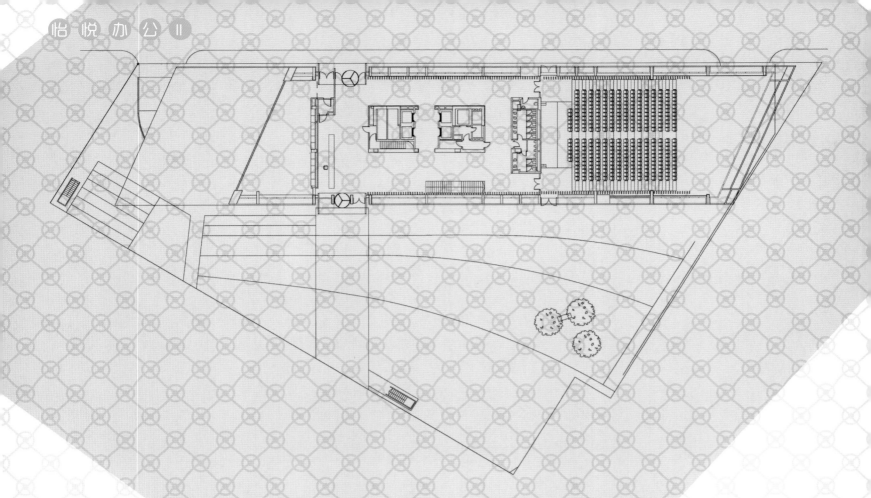

A.M.A. HEADQUARTERS

A.M.A. 总部办公楼

Rafael de La-Hoz Arquitectos

建筑机构：Rafael de La-Hoz 建筑事务所
建筑面积：10 000 平方米
项目地点：西班牙马德里市
摄　　影：Duccio Malagamba, Javier Ortega

Architects: Rafael de La-Hoz Arquitectos
Area: 10,000 m²
Location: Madrid, Spain
Photography: Duccio Malagamba, Javier Ortega

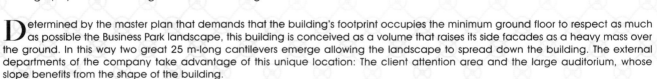

Determined by the master plan that demands that the building's footprint occupies the minimum ground floor to respect as much as possible the Business Park landscape, this building is conceived as a volume that raises its side facades as a heavy mass over the ground. In this way two great 25 m-long cantilevers emerge allowing the landscape to spread down the building. The external departments of the company take advantage of this unique location: The client attention area and the large auditorium, whose slope benefits from the shape of the building.

The facade design clearly presents the solution of the structural difficulties. In this way, a game of tension elements placed diagonally is sketched on the glass facade which directly reflects the law of the strains of the structure. Located at the Cristalia Business Park in the "Campo de las Naciones" of Madrid, the 10,000 m² outstanding building has seven office floors assigned to be the headquarters of an insurance company.

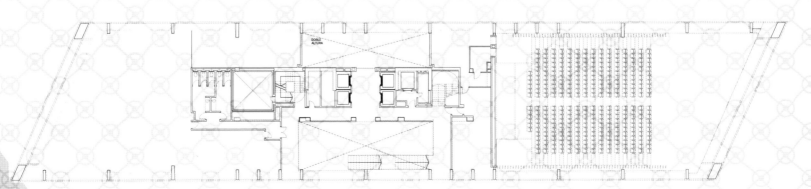

总体规划要求，建筑物需尽可能地占据最小的地面空间，为商业公园的景观提供更开阔的空间。因此本案设计为一侧立面高抬、悬于地面之上的建筑，这样便形成了两条 25 米长的悬臂结构，确保景观沿着建筑分布。该公司的外务部门利用区域特点，设置客户游览区域和一个大型礼堂，两区域的倾斜面直接采用建筑外墙。

外墙设计解决了结构方面的难题，一系列的张力元素以对角线形式置于玻璃立面上，直接反映了结构的张力定律。这座雄伟的建筑位于马德里的 Cristalia 商业公园，占地 1 万平方米，有 7 个办公楼层，现已作为一家保险公司的总部。

SUGAMO SHINKIN BANK / SHIMURA BRANCH

日本东京巢鸭信用银行志村分行

Emmanuelle Moureaux
Architecture + Design

设 计 师：Emmanuelle Moureaux
空间设计：Emmanuelle Moureaux Architecture + Design
客　　户：Sugamo Shinkin Bank
建筑面积：762.53 平方米
项目地点：日本东京市
摄　　影：Nacasa & Partners Inc.

Architect: Emmanuelle Moureaux
Space Design: Emmanuelle Moureaux Architecture + Design
Client: Sugamo Shinkin Bank
Area: 762.53 m²
Location: Tokyo, Japan
Photography: Nacasa & Partners Inc.

Rainbow Mille-feuille

Sugamo Shinkin Bank is a credit union that strives to provide first-rate hospitality to its customers in accordance with its motto: "we take pleasure in serving happy customers." Having completed the design for branch outlets of Sugamo Shinkin Bank located in Tokiwadai and Niiza, we were also commissioned to handle the architectural and interior design for its newly rebuilt branch in Shimura. For this project, we sought to create a refreshing atmosphere with a palpable sense of nature based on an open sky motif.

12 Layers of Color

A rainbow-like stack of colored layers, peeking out from the facade welcomes visitors. Reflected onto the white surface, these colors leave a faint trace over it, creating a warm, gentle feeling. At night, the colored layers are faintly illuminated. The illumination varies according to the season and time of day, conjuring up myriad landscapes.

A Piece of The Sky

Upon entering the building, three elliptical skylights bathe the interior in a soft light. Visitors spontaneously look up to see a cut-out piece of the sky that invites them to gaze languidly at it. The open sky and sensation of openness prompts you to take deep breaths, refreshing your body from within.

Fuzzy Puffs

The ceiling is adorned with dandelion puff motifs that seem to float and drift through the air. In Europe, there is a long and cherished custom of blowing on one of these fuzzy balls while secretly making a wish. Bits of fluffy down gently dance and frolic in the air, carried by the wind. ATMs, windows, consultation booths and an open space laid out with chairs in 14 different colors are located on the first floor. The second storey houses offices, meeting rooms and a cafeteria, while the third floor is reserved for the staff changing rooms. Three long glass airwells thread through the first and second levels of the building, flooding the interior with natural light as well as "blowing" air through it.

彩虹千层糕

巢鸭信用银行是一个信用合作社,其宗旨是"以让顾客满意为荣",努力为顾客提供一流的服务。在完成了巢鸭信用银行常盘台和新座市的两个分行的设计后,建筑师又接受委托,负责刚刚重建的志村分行的建筑与室内设计。在这个项目中,建筑师集中营造一种清新的氛围,以开放的天空主题为主调令人感知大自然。

12层色彩

彩虹色的重叠层,探出外立面,迎接往来客人。颜色反射到白色表面,留下淡淡的色氲,添上温馨柔和之感。彩色层在晚上会发出点点微光,随着时间和季节的更替营造出无数胜景。

顶空

进入大楼后,顶空的3个椭圆采光井为室内提供了柔和的自然光线,使得客人不由自主地向上张望,并自得其乐地欣赏。开阔的天空及开放的氛围让人不禁深呼吸,将清新的感觉带回体内。

蒲公英毛球

天花板上装饰着蒲公英主题的图案,给人蒲公英漂浮在空中之感。欧洲有一个极受重视的悠久传统,即在吹蒲公英毛球的同时可以默默地许愿,看点点蒲公英随风荡漾,在空中缓缓地飞舞下落。建筑1层设有自动提款机、窗口、咨询台及一个内置了14种颜色座椅的开放空间。2层设置办公室、会议室和一间自助餐厅,3层则是员工们的更衣室。长长的玻璃采光井穿透建筑的1层和2层,为这两个楼层带来自然光和习习微风。

ARCHITECTURE OFFICE——LOFT
LOFT 建筑办公室

NMH

设 计 师：Natalia Figurska, Moncho González Berenguer,
　　　　　Hugo González Berenguer
建筑面积：140 平方米
项目地点：西班牙瓦伦西亚市

Architects: Natalia Figurska, Moncho González Berenguer,
　　　　　　Hugo González Berenguer
Area: 140 m²
Location: Valencia, Spain

The main idea of the project is to maintain the space as it is, a volume of 5 m×11 m×11 m. It includes a bathroom and a kitchen area in one piece. White is used as the base color and green as the soloist, characteristic of the logo and the study's philosophy in its quest for integration of nature with architecture. The space is set up temporarily in two parts, a work area, and a relaxation and meeting area. The table that dominates the relax-meeting space also has this double character, is a table to meet but having the dimensions of a ping-pong one, putting on a net serves to relax for a while. What's more, the meetings are always a give and take between clients and professionals, reminiscent of a game of ping-pong. The only built element is covered with artificial grass. All installations are hidden in the technical floor, except for the air conditioning that is hidden in the grass. With a glass surface of 110 m² divided between the south and west facades, there are no lighting needs, maybe just protection by the mechanical sun protection louvers. In winter months, after office hours some additional lighting is necessary, and it is produced throughout the space overhead, there also can be found various points of light in the workplace.

此案计划在原有的 5 米 ×11 米 ×11 米的空间中打造一个办公区。为迎合公司标识语的独创特色，使用白、绿作为主色，象征着自然与建筑的结合。原空间为带有一厨一卫的单元房，现在暂被重整为两部分——工作区和休闲区、会议区。桌子是休闲区的主体，具有双重用途，既可作为会议桌也可闲暇时布置上网纱作为乒乓球桌。事实上，与客户的会谈就如同桌上乒乓，你来我往。空间中唯一设有屋顶的区域覆有人造草皮，所有装置都被隐藏在技术地板中，空调则被遮蔽在草丛中。110 平方米的玻璃外墙被分隔为南边与西边，光线充足，有时甚至需要机械百叶窗"防晒"。除了冬季下班时，会用到置于顶部的多种辅助光源设备。

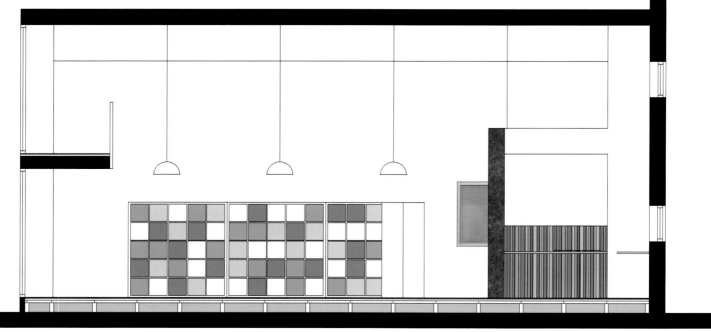
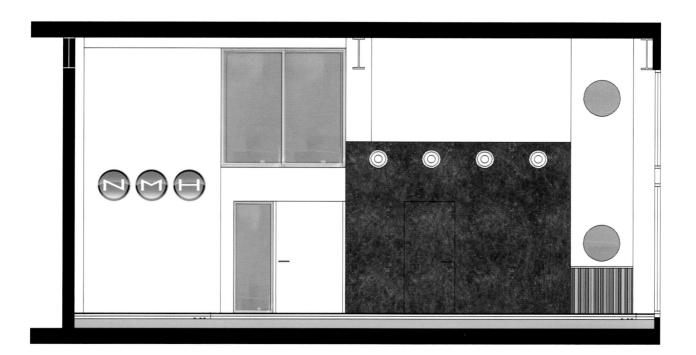

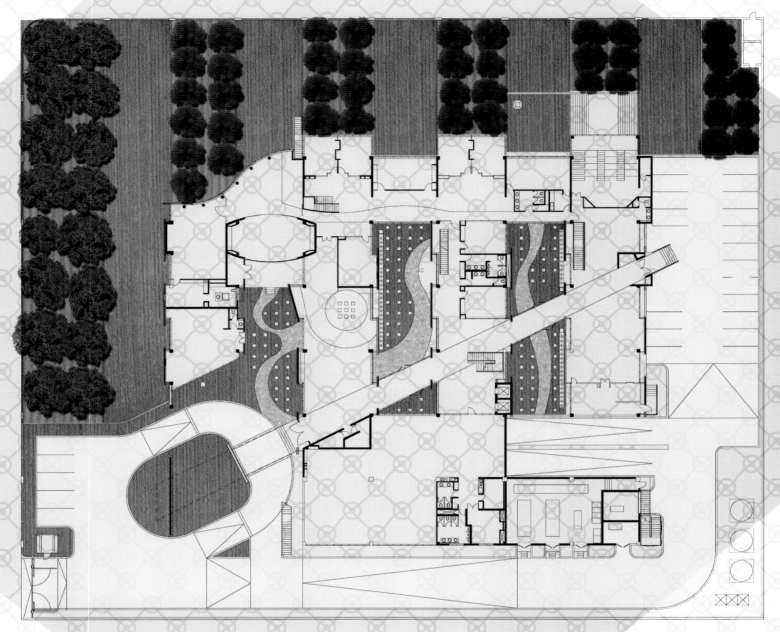

CORPORATE OFFICE FOR APOLLO TYRES

阿波罗轮胎公司办公室

Morphogenesis

室内设计：Morphogenesis
客　　户：Apollo Tyres Ltd.
建筑面积：9 000 平方米
项目地点：印度古尔冈市
结　　构：Mehro Consultants

Interior Design: Morphogenesis
Client: Apollo Tyres Ltd.
Area: 9,000 m²
Location: Gurgaon, India
Structure: Mehro Consultants

The corporate headquarters of the Apollo Tyres Group is an instrument to address issues that are central to the rapidly evolving Indian workplace. The approach was to ensure that the project sits within the conceptual, functional and aesthetic framework of a Contemporary Corporate Design Agenda. The site is a vast expanse of an empty field in the middle of nowhere. Flat and lacking any topographical character, it was extremely difficult to contextualize an iconic corporate office building that the design briefly specified. This helped to generate a morphology that goes against the typical corporate office as a building type in India; that of a sealed box, within which the office environment is predictably placed.

The notion of a mini-urban center was conceived conceptually, which would have a texture and fabric similar to that of the older cities such as Jaipur or Shajahanabad. Devoid of any context, a fragmented morphology was hence developed, which helped to generate the conceptual intent of the office campus being a "micro city". 9 meter wide parallel linear strips running SE-NW are grafted across the site. These help to create the main programmatic zone divisions into Nature zone, Work zone, Social Zone, Recreation zone, Transaction Zone, Chance Event Zone, Instructional Zone, etc. which provide an interface and permeability between the striations. Courtyards and terraces are provided right up to the second floor level with various linkages in the form of steel bridges in order to bridge the resultant diverse fragments. A movement spine traversing the programmatic striations stimulates the activity zones and restructures the office space.

This morphology further helps to articulate the thermal banking and day lighting strategy. Internal and External courtyards abut the striations and terraces are provided to bring in maximum daylight. The intention was to create a virtually blinds-free environment. The terrace gardens also provide a high level of thermal insulation. A staggered skin treatment further enhances this function; the skin is developed as 2 sets of planes (solid and glass)—the solid set of planes is staggered to shade the glass surfaces to give protection from the sun and keep the glare out. This also ensures that solar ingress is limited to the winter months. The resultant of this strategy is an air-conditioning system which averages about 23 m^2 per tonne of air conditioning which is far better than the industry standards.

An integrated building management system is employed to link all services to human occupancy. This is done to monitor entries/exits through three main access points in order to sense where each occupant is in the building and thereby, activating all services such as air conditioning, electrical systems and Lighting based on occupancy. The building is preset to a standby mode wherein all these services are operational only in a limited capacity. On an occupant entering the building, the proximity readers sense the activity and shift all zones to occupied zone where all the services work in full capacity and the system works vice versa on the occupant leaving the building. This leads to the creation of pre-cooled environments and adoption of a skeleton lighting procedure as and when required. Hostelling further adds to the flexibility of this system and permits dynamic requisitioning of the work spaces. Also, it further helps to achieve an interactive control of the facility management system and complete energy efficiency in a well integrated environment.

The architecture is a consequence of an "arts and crafts approach". Varied and tectonic use of material extends through out and infests itself dramatically in the sculptural external fire escape staircase in rippling stainless steel. The building facade itself is fragmented, a series of broken angular planes in glass and aluminum. The entrance is a curved reflective glass wall forming an entrance court dominated by a stainless steel screen wall and the dramatic porch jutting out from the building. The building has no formal entrance and the distinction between the interior and exterior space is lost in the morphology of the building. The design of the Apollo Tyres Corporate Office takes into consideration the importance and relevance of energy conscious design within the modern work culture.

阿波罗集团的公司总部是解决印度工厂快速发展问题的核心建筑。其方法在于确保项目始终顺应着当代企业设计议程，使其框架结构独具概念性、功能性以及美感。工厂选址在偏僻广袤的空旷处。此处地势平坦，不具任何地形特征，因此要将一座外观明确化、具有标志性的企业办公楼融入此处似乎有些困难。但此举有助于打破印度传统的办公建筑模式，不再将办公环境框入密闭空间内，开启形态学的创新之门。

迷你城市这一概念的孕育，从本质和构造上来说，与斋蒲尔、巴德这些古老的城市的兴起有相似之处。分散形态不受环境约束得以发展，这将促进办公社区向"微城市"的转化。9米宽的平行线条呈西北—东南走向贯穿整个空间，据此分割主功能区。区域大致分为自然区、工作区、社会区、休闲区、交易区、偶发事件区和教学区等，它们彼此独立却又具相通性。庭院和露台位于第二层，以类似钢桥的连杆结构将零散的部分加以衔接。一个活动立杆横穿整个功能区，丰富了活动区域，同时也重整了办公空间。

进一步来说，这种形态利于热量储备和日光照明，能够最大限度地为邻近各区的露台和内外庭院提供自然光，创造了一个几近通亮的空间环境。露台花园装有一个高质保暖的热隔板，交错的表层纹理进一步提升其功能性。改进外层，安装两套板层（固体和玻璃），固体板层交错掩盖玻璃板层，防晒同时遮挡刺眼的光线。这也保证了在冬月阳光的射入。该策略综合结果就是，空调系统调节在平均每23平方米一吨的空气流量，这远远优于工业标准。

一个完整的建筑管理系统可以确保所有服务端口为人所用。其目的是通过三个主要接入点来监控出入口，以定位楼内的每个成员，根据人员位置激活空调、电力系统和光照等所有设施。建筑内设施预设为待机模式，只有在特定情况下才可以启动。当一个人进入到这座建筑中，邻近的感应器会定位其活动范围，将所有区域的服务器转移到该指定区域，并全容量运作；当此人离开该建筑时，系统运行反之亦然。这将有助于环境预冷并营造人们所需的照明环境。旅馆式办公形式进一步增添了系统的灵活性，满足办公空间动态化的需求。同时，这也有助设备管理系统实现交互式管理，在集成环境中增强能源利用效率。

该建筑架构是"艺术和工艺手法相结合"的产物,建筑内部整体使用各式构造材料,外部也设有雕塑波纹状的不锈钢逃生梯,颇为引人注目。这座建筑的立面采用碎角纹玻璃和铝材,构成形式较为零散。入口通道镶有弧形的反光玻璃墙,设置不锈钢屏幕和高出建筑的醒目门廊。建筑不设正式入口,室内和室外不存在建筑形态差异。阿波罗轮胎公司办公楼的设计将能源意识和现代工作文化相结合作为重点考虑。

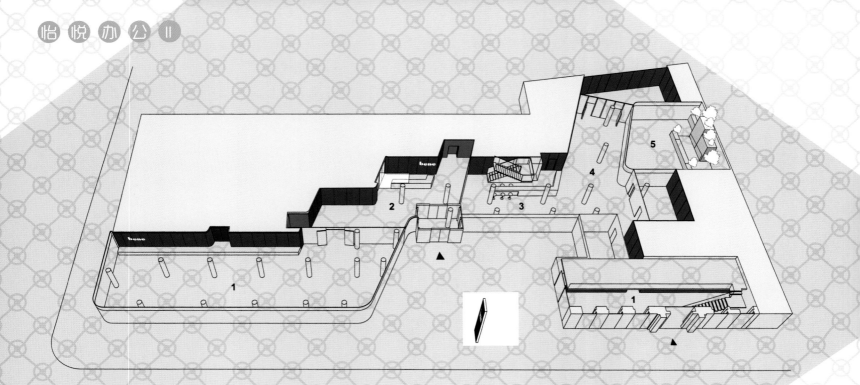

BENE FLAGSHIPSTORE
Bene 旗舰店

SOLID Architecture ZT GmbH

设 计 师：SOLID Architecture ZT GmbH
客　　户：Bene AG
建筑面积：950 平方米
项目地点：奥地利维也纳市
摄　　影：Günter Kresser

Architects: SOLID Architecture ZT GmbH
Client: Bene AG
Area: 950 m²
Location: Vienna, Austria
Photography: Günter Kresser

SOLID Architecture was commissioned to design the showroom of office furniture manufacturer Bene in a newly erected building at No. 4–8 Neutorgasse in Vienna. With a floor area of 1,000 m² the showroom is located at ground floor level and forms a band that runs along the facade. Bene also uses the first floor and a part of the second floor as offices. The intention was to offer visitors an interestingly varied tour through the world of Bene and at the same time to create a place with its own identity and an unmistakable character.
Design of the showroom
The walls, floors and ceilings are the surfaces that determine the character of the space: For each of them a single material was used throughout. To accommodate the different spatial functions and presentations of the show room, these design elements can be adapted as required.
Wall cladding—the "backbone"
The entire rear wall of the showroom is clad with panels covered in fabric and forms its spatial backbone. The panels are built up on a modular system so that they can be easily exchanged and interchanged. The backbone defines and delineates the space. In the area that is also used for events the panels are backed with an additional sound-damping acoustic inlay.
Ceiling
The acoustic ceiling to the showroom is made of granulated blown glass. All the linear light fittings and technical services in this ceiling run parallel to Neutorgasse. This direction is continued throughout the entire showroom. All the inserted services are flush with the ceiling.
Flooring—terrazzo
The entire showroom, including the service spaces, has a light-colored terrazzo floor. It forms a continuous surface on which individual functions and presentations can be individually emphasized by means of carpets.
Color scheme
In selecting the materials for the floor and walls it was important that these surfaces should provide a harmonious background for furniture with different surface finishes and colors. White surfaces as well as wood finishes with a strong grain should be able to make an impact and contrast with the setting. At the same time the showroom was to convey a warm and friendly underlying mood. Consequently it was decided not to use monochrome surfaces in the design of the showrooms but materials made up of several colors that also have a high haptic quality. The large area of light-colored terrazzo has a white, grey and brown natural stone "grain". The dark brown fabric used to cover the "backbone" also follows the same concept; the material is woven from brown and black threads.
Internal staircase
The core of the Bene premises is a centrally positioned internal staircase that connects all the floors used by Bene. This stair forms the central communication area to which certain gastronomic functions are attached: the bar with the "office" on the ground floor, and the tea-kitchens and the areas for informal communication on the two upper floors. The internal staircase follows the design concept of the ground floor showroom. The staircase walls are clad with panels covered with brown fabric. In addition, a "cloud" of square LED lights extends from the ground floor to the second floor.

1 Multipurpose room
2 Conference room
3 Storage
4 Atrium

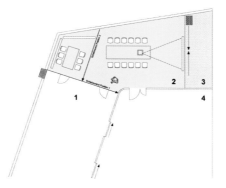

Two conference rooms

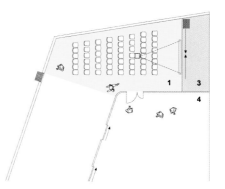

One function room

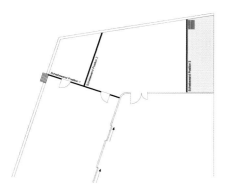

Mobile partition walls

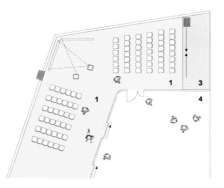

One large function room

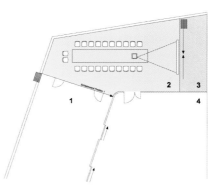

One conference room

SOLID 建筑事务所受奥地利办公家具公司 Bene 的委任,为其设计一个展示大厅。展厅位于维也纳 Neutorgasse 街 4 至 8 号新建的大楼首层,占地约 1 000 平方米,像一条随建筑外结构延伸的彩带。二层和三层的部分区域作为 Bene 的办公空间。经过此番设计,游客将会看到一条生动多样并充分展现 Bene 个性世界的道路。

展示空间设计
展示空间运用墙面、地面、天花板材质来突显空间特点,考虑到整个空间频繁使用多种材料,这些设计元素必须适应不同的空间功能需要,根据展示空间的特点做材质调整。

墙面——主背景
展示空间的所有背墙统一采用织物饰面板作为空间主要背景,面板采用模块化铺排,便于灵活地进行局部更换,墙面背景也以这种方式定位了空间格调。这种具备吸音功能的材料也被运用于活动区

吸音天花板采用粗糙的吹制玻璃材质。在 Neutorgasse 大楼里,天花板上所有线性内嵌灯具和技术设备平行排开,覆盖整个展示空间,所有嵌入设备与天花板保持同一水平面。

地板——水磨石
整个展示空间(包括公共空间在内)的地面都采用浅白色水磨石,形成连续统一的肌理,其中个别功能展区使用了小面积地毯加以区分。

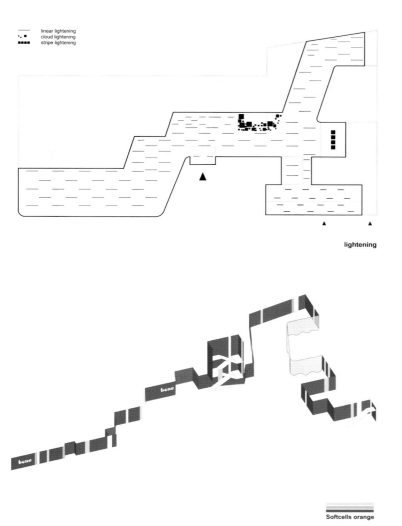

lightening

Softcells orange

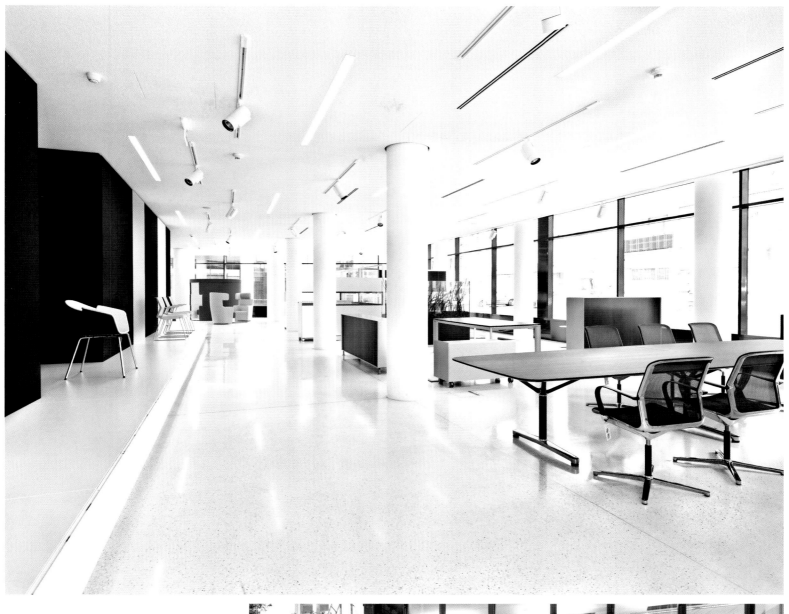

色彩搭配

地板和墙面的选材十分重要，鉴于这些所选材质表面色彩需与家具和谐统一，而不同家具有着不同的表面磨光和色彩。家具的白色表面以及木质纹理与背景颜色产生强烈对比与视觉冲击力，同时也为展示空间营造出一种友好而热情的气氛。因此，整个展示空间以不使用绝对纯色为原则，采用多种混合颜色的高品质材料。比如大面积的浅色水磨石是由白色、灰色和棕色天然石材混合打磨而成，墙面黑褐色织物饰面板也是由棕色与黑色原料混合编织而成。

内部楼梯

在建筑的中心，内部楼梯作为沟通的重要区域连接着所有楼层，同时也连通各个餐饮功能区，例如首层的吧台，其上两层的茶水区及非正式的交流场所。内部楼梯沿用与首层展示区相同的设计元素，包括墙面的黑褐色织物饰面板以及在吧台出现过的"cloud"方形 LED 灯。

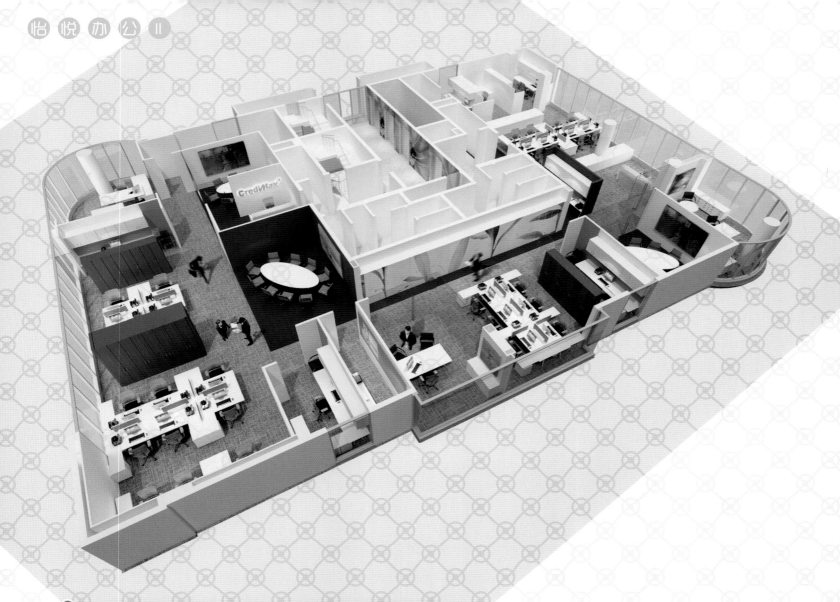

CREDIMAX HEAD OFFICE

CrediMax 办公总部

dwp

建筑机构：dwp
建筑面积：2 500 平方米
项目地点：巴林泽夫市

Architects: dwp
Area: 2,500 m²
Location: Seef, Bahrain

World-class architecture and interior design firm dwp was requested to design the new bank and office headquarters for CrediMax, in Seef Bahrain. Set over 5 floors, the space comprised a total of 2,500 m² that CrediMax wanted to reflect the company image and spirit. The design briefly stipulated a young, modern and internationally styled interior, without using wood finishes, but that subtly reflected the local Arabic culture. For the headquarters, dwp's design intent was to individually color each floor, thus giving each department its own identity. Furthermore, dwp commissioned a local Arab artist to paint 15 sizeable canvases, incorporating each color in an abstract Arab-styled scene.

For the banking hall, dwp designed a monochrome space, accentuated by large colored graphic walls, to allow for the combination of BBK and CrediMax banking, which share the space, yet have separate corporate identities.

世界级建筑及室内设计事务所 dwp 负责设计 CrediMax 在巴林泽夫市的新银行总部办公室。办公空间超过 5 层楼，总面积为 2 500 平方米，CrediMax 计划以其展现公司形象和企业精神。室内设计以"年轻、现代、国际风格"为简旨，不采用木饰面而巧妙地反映出阿拉伯当地文化。在总部办公室方面，dwp 为每层楼配上各自的颜色，从而赋予每个部门各自的特性。此外，dwp 还委托当地的阿拉伯艺术家绘制了 15 幅大型油画，将每层的主体颜色融入到该处抽象的阿拉伯风格场景中。

dwp 将银行大厅设计为一个单色空间，利用大型彩绘墙突出设计效果，让共用办公空间的 BBK 和 CrediMax 银行在和谐中突显各自的与众不同。

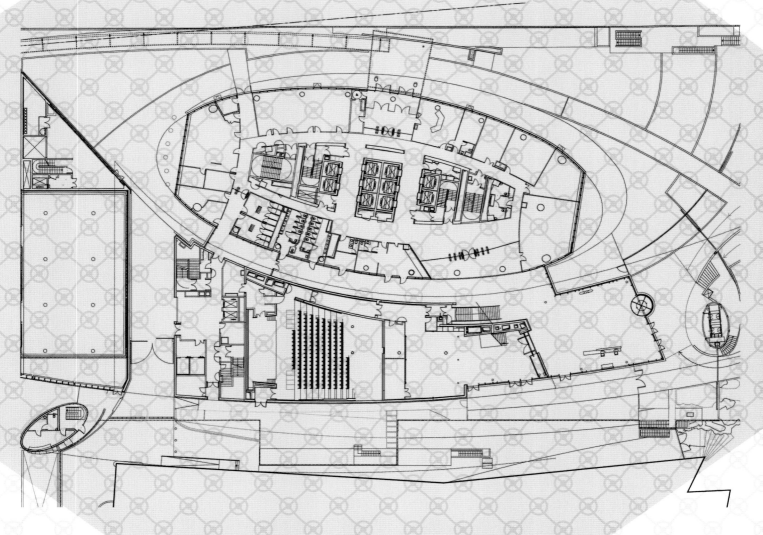

BOUYGUES TELCOM TOWER
布依格电信办公总部

Arquitectonica, Wilmotte & Associés SA

建筑公司：Arquitectonica
客　　户：Bouygues Immobilier
建筑面积：32 000 平方米
项目地点：法国巴黎市
摄　　影：Paul Maurer, Eric Morrill

Architects: Arquitectonica
Client: Bouygues Immobilier
Area: 32,000 m²
Location: Paris, France
Photography: Paul Maurer, Eric Morrill

The project is organized around the power of a tower located at the center of the composition. It rises in the same location as the existing EDF tower, yet expands its footprint and height to maximize the volume within permitted envelope.
The elliptical form gives the form direction and multiple frontalities. There is no front or back, yet its directional quality conveys movement from Issy to Paris, from the streets to the peripherique, from land to river. Its soft-shaped edges avoid hard edges and corners, implying a democratic space and certain sensuality in its relationship to the green landscape that surrounds the building at its base.
The elliptical prism is eroded as it reaches the sky, following the angles of the envelope. This erosion results in two of angled planes that reveal the curved plan in elevation. The receding floor plates provide additional space that can be captured for functional use. They also introduce executive floors with smaller footprints and angled glass surfaces that act as greenhouses and add a uniqueness to the top floor spaces.
The glass skin of the tower is articulated by a series of recesses that imply a rotational force. Like speed lines or gills, they also provide occasional terraces and corner offices at selected locations throughout the building and add depth to the otherwise pure prismatic volume.

The tower is organized around a center core of elevators, stairs, technical rooms and toilets. The balance of the center space is dedicated to meeting rooms and office to support functions. The elliptical form allows maximizing the amount of office space by widening the center to accommodate the core and narrowing towards the ends to minimize interior space. The result is a large floor plate area with maximized perimeter offices within the permitted 75 m maximum length.
The site also contains a low-rise building that accommodates the balance of the permitted buildable area. A bar is placed along the street running parallel to the ellipse. It bends to form a sinuous S curve that responds to the force of the ellipse. Its plasticity adapts to the sensuality of the ellipse. The S bar contains more classic offices either at 18 m in width or two zones of 12 m along a linear glass spine. A third building exists across the Paris boundary.

这个项目以地块正中央的大厦布局设计而展开。它与既存的 EDF 大厦处于同一地段，但又在可行的范围内扩大面积与提升高度，以达到最大化容积的目的。

椭圆外形予以建筑物方向感和全方位正面感。它并无前后之分，但其方向感却很好地表现了从伊希（法国城市）到巴黎、从街道到外围和从陆地到河流的动态变化。柔和的边缘设计避免出现尖锐突起或棱角，暗示着空间的民主性，彰显着其与建筑周围绿色景观的完美融合。

椭圆的棱柱体直指天空，依照角度形成外围包络。这种处理手法产生了两种不同角度的平面，建筑从正立面扩展到曲面区。下沉的楼板为各种功能用途提供附加空间。面积略小、采用弯角玻璃面设计的办公楼层被用作建筑物的温室，为顶楼增加了独一无二的特定感。

大厦的玻璃幕墙由连续的凹槽依靠旋转力铰接而成，这种设计模仿鱼的流水线和鳃部，为空间提供了临时门廊和角落办公室，增加了原棱柱形空间的深度。

大厦中部区域设有电梯、楼梯、机房和卫生间，空间布局均衡，使得会议室和办公室功运转正常。椭圆外形设计增加中部容量来满足核心需求，收窄两端以最小化内部空间，从而使得办公空间得到最大限度的扩展。最终在 75 米允许长度范围内，形成这个地基面积庞大，周长最大化的办公空间。

此建筑设计还包括一栋低层建筑，用来维持各建筑区域平衡。沿街放置的条形建筑与椭圆外形的建筑同向，并被弯成"S"形以衬托椭圆的力道，而其可塑性又与椭圆的魅感相呼应。S 形建筑内部沿其玻璃大梁开辟了更多布局经典、宽度为 18 米或 12 米的办公室，它也成为又一座横穿巴黎边界的建筑。

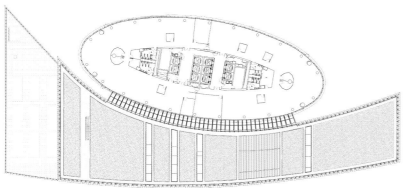
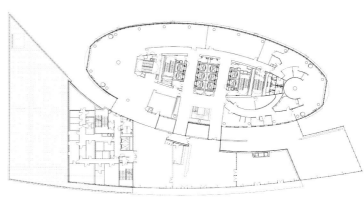

FLEXIBLE SOLUTIONS FOR DESIGN OFFICES

Design Offices 的活性办公空间设计

Dan Pearlman
Marken Architektur GmbH

设 计 师：Marcus Fischer
客　　户：Design Offices GmbH
建筑面积：1 500 平方米
项目地点：德国科隆市
摄　　影：Agnes Doroszewicz

Architect: Marcus Fischer
Client: Design Offices GmbH
Area: 1,500 m²
Location: Cologne, Germany
Photography: Agnes Doroszewicz

Stylish, modular, and not just eye-catching—this office furniture can do it all and even more. The lines we developed exclusively for Design Offices take into account ergonomic considerations such as adjustable table heights or usage aspects where work settings for communicating while standing are required. By removing shelf or lateral elements, privacy and/or additional communications and storage areas are created.

Despite its superior design, price points for the furniture are considerably lower than comparative models. With very reduced and simple forms, they provide solutions to all daily office demands and their feel-good character satisfies emotional and aesthetic requirements. Since January 2010 they are on display at the Cologne showroom of Design Offices.

这套办公家具不单只是时尚、有序、迷人，简直无与伦比。简约优雅的线条由 Design Offices 依照人体工程学独家设计，桌子根据工作上的沟通需要来调节站立或坐立的高度。通过移动置物架与侧边元素，创造出其他私密沟通空间与储物空间。

这套办公家具设计优越精湛，但价格却比同类产品实惠许多。简洁素净的造型满足了日常办公的一切需求，令人怡悦的特质也满足了现代人对办公环境的感性美学追求。自 2010 年 1 月起，这套办公家具在 Design Offices 的科隆展厅展出。

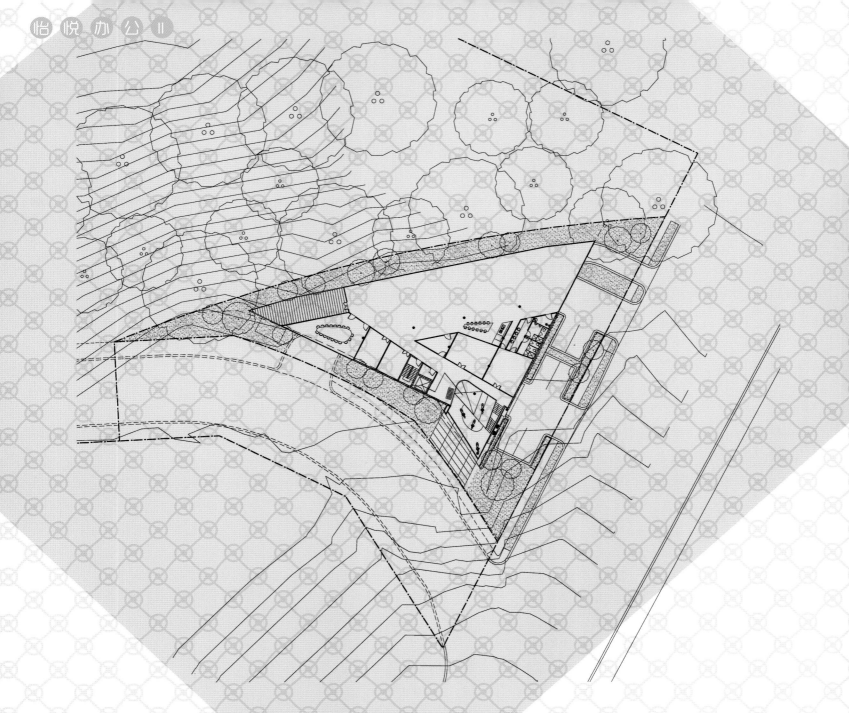

NEW HEADQUARTERS FOR HARLEY DAVIDSON

Harley Davidson 新总部大楼

Tony Owen NDM Architects

建筑面积：3 000 平方米
项目地点：澳大利亚悉尼市

Area: 3,000 m²
Location: Sydney, Australia

Sydney Architects Tony Owen NDM has designed the new Australian headquarters Harley Davidson. The building is located in Lane Cove and forms an iconic gateway to the new Lane Cove River business park.
We sought to design a building that reflects the uniqueness of Harley Davidson. HD is not simply a brand, for many people it is an entire lifestyle and attitude. HD has a unique philosophy; it is at once about the expression of function and beauty through pure design, but it is also about freedom; the freedom of self expression and the freedom of the open road. We could relate to this image, it is not only about good design, but also about challenging the norm. We decided to design a building that expressed this freedom and speed.
For design inspiration we looked to the bikes themselves; their emotion and efficiency. The geometry of the engines forks and frames can be seen in the lines of the building. The building does not copy them, however, it suggests this movement and style. Rather than use the shape literally, we sought to express the elegance and aerodynamics of this movement in the lines of the building.

The brief for the building was a strong reflection of the Harley Davidson culture. We gave as much emphasis to the gymnasium and break-out areas as the office and storage space. We designed the building to reflect this. We located of all of the recreational and break-out areas near the entry. You enter into a central mezzanine. From there you can see all of the areas that reflect the Harley lifestyle: the showroom, cafe, library, even the Gym. You are immediately aware of what Harley Davidson is all about. We designed the building so that, from the entry, you can also look down into the technical workshops and training areas. In this way you get a sense of the technical aspects of the company. It was important not to lose sight of the grungy side of the motor cycles as well.

We utilized 3D modelling software in the earliest stages of the design. The office is increasingly using 3D modelling and parametric tools both in the generation of formal solutions and in the development and construction of complex geometries. For Harley we experimented with different geometries to understand the spatial possibilities of our ideas and also how the forces working on the site influenced these movements. Later we worked with the engineers to model every component of the main space. Because the geometry is complex, it was important to know how the structure interacted with the metal cladding. By modelling every element of structure and facade we knew exactly how each piece related to another in space. This technology makes building a structure such as this quite straight forward and much easier than it might have been in the past.

The facility contains administrative offices, technical training and storage facilities for the iconic motorcycle company. The landmark building will form the striking centerpiece for a new high-tech business park on the Lane Cove River, currently being developed by Demian Developments Pty Ltd.

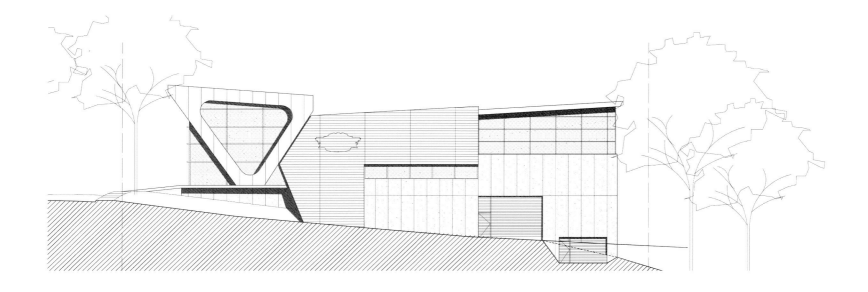

悉尼建筑机构 Tony Owen NDM 为 Harley Davidson 公司设计了位于澳大利亚的新总部。该总部位于悉尼市西北城区——兰蔻，是通往兰蔻河商务园的一座标志性建筑。

建筑师计划设计一座能够体现 Harley Davidson 独特感的建筑。HD 不单是一个品牌，对于许多人来说它代表了整体的生活方式与态度。HD 的独特理念不仅在于设计所形成的功能与美，更关乎自由——自我表达的自由与任意驰骋的自由。所以 HD 新总部大楼的设计须是卓越的、反传统的，建筑师应在这座建筑中体现出自由感与速度感。

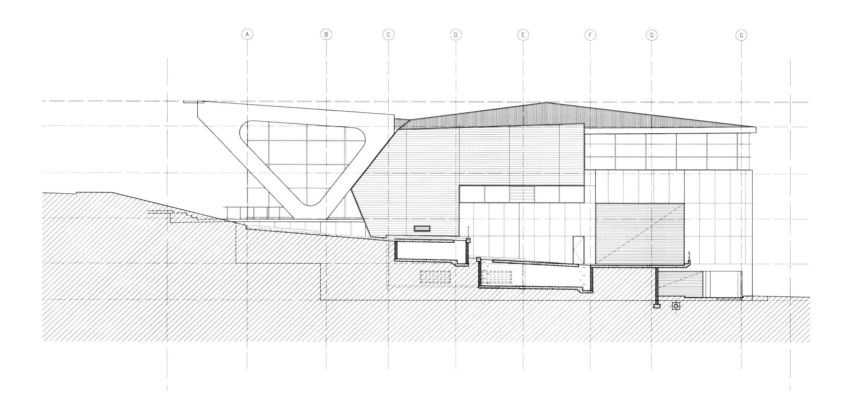

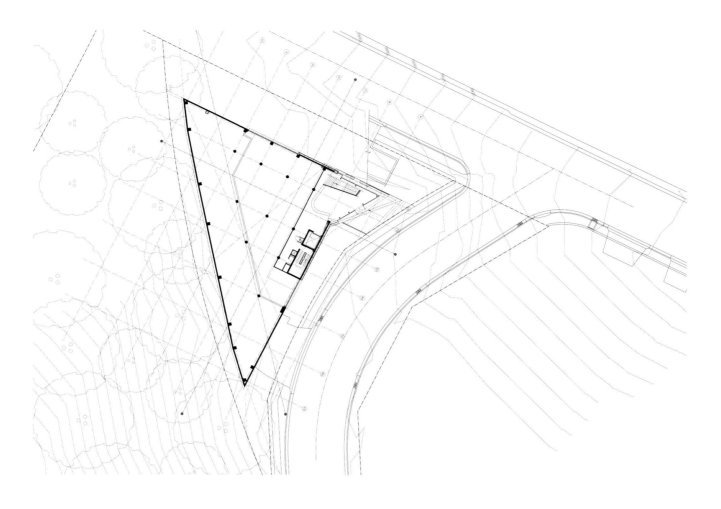

建筑师从自行车的感性与效率方面获得了设计灵感。建筑的线条展现出发动机外观框架的几何结构，没有生硬地模仿，而是蕴含了其动感与风格，没有机械地将原结构全盘套用，而是在建筑中表现出建筑线条的优雅与这一运动所体现的空气动力学。

这个设计要旨强烈反映了 Harley Davidson 的文化，关注办公室与储物空间的同时，建筑师也没有忽略健身房与休息空间的设计。从建筑中可以看出这点：消遣区和休区都被布置于入口处附近。从中央夹层可以看见展厅、咖啡厅、图书馆，甚至健身房。各个地带都能反映出 Harley 的生活方式，从中人们一眼就可领略到 Harley Davidson 的内在品质。人们可以从入口处俯看下方的技术车间与培训区，甚至看清摩托车的污漆面，进而了解该公司的技术水准。

建筑师在早期的设计中应用了 3D 建模软件，也在确定标准方案和构建复杂几何结构时频繁地使用 3D 软件和参数工具。建筑师为 Harley 尝试了几种不同的几何结构，从而能够更好地确定方案的实用性及建筑本身的实际呈现将会受到怎样的影响。此后建筑师与工程师共同为空间中的每个运作元件塑模。由于几何结构相当复杂，建筑师须理解整体架构与金属板的磨合运作。经过对内部与外部结构的每个元素进行塑模，最终对空间中每个元素之间的关系了如指掌，现代技术使建筑结构比过去更容易地呈现出独树一帜的姿态。

办公设施包含了行政办公室、技术培训部及摩托车公司标志性的储物设施。这座地标性建筑由德米安发展私人有限公司开发，将会成为兰蔻河上新兴高科技商业园里引人注目的核心建筑。

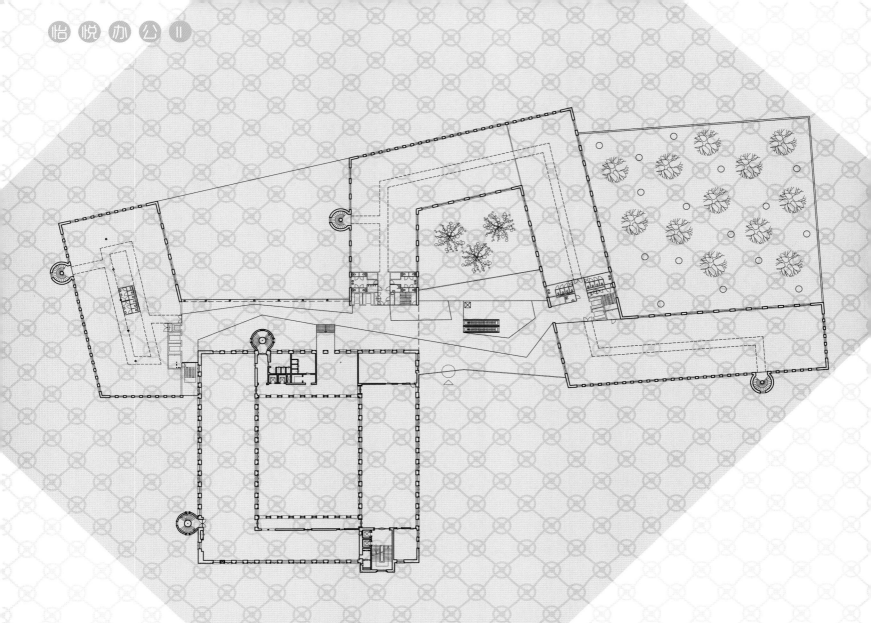

ESSENT HEAD OFFICE

ESSENT 办公大厦

Pi de Bruijn, de Architekten Cie.

室内设计：V. Ulrich, C. Fuchs, A. Icinsel
项目团队：T. Berkhout, O. Sarafopoulos, A. van der Hoek, M. van Aalderen,
　　　　　Th. Martens, R. Garritsen, A. van Gelderen, P. Kinfu, J. Kriene,
　　　　　G. Vissers, A. Janson, L. de Jong
客　　户：Essent Facilities
建筑面积：36 000 平方米
项目地点：荷兰登波士市

Interior Design: V. Ulrich, C. Fuchs, A. Icinsel
Project Team: T. Berkhout, O. Sarafopoulos, A. van der Hoek, M. van Aalderen,
　　　　　　　Th. Martens, R. Garritsen, A. van Gelderen, P. Kinfu, J. Kriene,
　　　　　　　G. Vissers, A. Janson, L. de Jong
Client: Essent Facilities
Area: 36, 000 m²
Location: Den Bosch, the Netherland

The former PNEM building (C. de Bever, 1956) was extended and renovated by adding new wings to the rear of the castle-like structure. Although this drastically altered the scale and organization of the complex, old and new still form a harmonious whole thanks to subtle similarities correspondences in scale, formal idiom and roof structure. A "high street" connects the wings with the original building and gives access to the office spaces and collective functions. The old courtyard has been transformed into a glass-covered atrium that serves as a foyer and restaurant for the conference rooms surrounding it.

建筑师将原来类似城堡结构的 PNEM 大厦 (C. de Bever, 1956) 翻新，在其后方加建了新的侧楼。虽然加建的建筑极大地改变了原有的规模和复杂的结构，但由于两者在比例上具有微妙的相似之处，新旧建筑仍能组成一个和谐的整体。一条"高街"将新翼与原有建筑连接起来，它除了作为进入办公空间的通道，还具有其他广泛的用途。原有园林景观也被改造成玻璃中庭，作为门厅大堂及为周围会议室提供餐厅服务的场所。

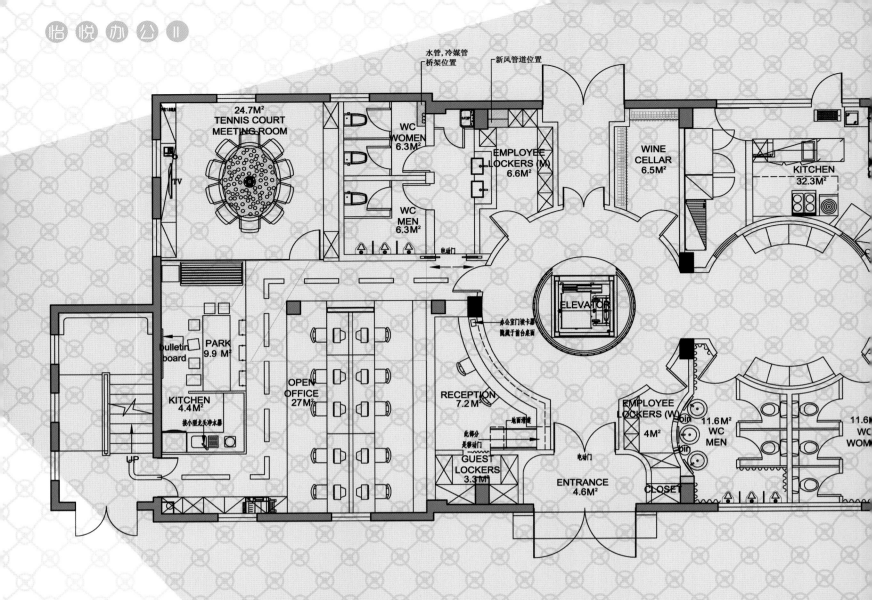

DUNMAI OFFICE

Dunmai 办公室

Dariel Studio

设 计 师：Thomas Dariel
建筑面积：1 200 平方米
项目地点：中国上海市
主要材料：真石漆，大理石，木地板，马赛克
摄　　影：Derryck Menere

Architect: Thomas Dariel
Area: 1,200 m²
Location: Shanghai, China
Main Materials: Nature Stone Paint, Marble, Wood Floor, Mosaic
Photography: Derryck Menere

The Dunmai office is located in the fashionable creative park, Happy Wharf, the South Bund, which was the Shanghai Happy Motorcycle factory in the before. The area in the Nanpu bridge block with heavy traffic is not so eye-catching. When the evening lights are lit, the lemon yellow lamp lighting casts light on the simple and unsophisticated skirt buildings and the continuous bamboo also reflects the well-spaced light shadow, which draws the outline of a quiet, leisurely, open creative space. The South Bund is the symbol of Shanghai and also an eternal feeling at the heart of the Shanghai people from the last century to nowadays such as the old wharf, outer road, sixteen shops... Only by seeing these place names, in the mind of the Shanghai natives the sounds of the waves come slowly and the siren sob of the large ship and the ferry starts in the dense morning fog. Today the South Bund not only carries forward the classical elegance and luxury of the Old Bund and the historical prosperity of the Dong Jiadu but also links the integrated engineering representing science and technology in the Bund and the Expo Park together.

The client of the project is a creative company for holding activities. After taking the characteristics and requirements of the client into consideration fully, the two designers Thomas Dariel and Benoit Arfeuillere have decided to build it into a creative office which conforms to the park concept and also brings a joy and relaxed working atmosphere for employees. They make use of the high-tech products to reflect the utility and functionality of the office.

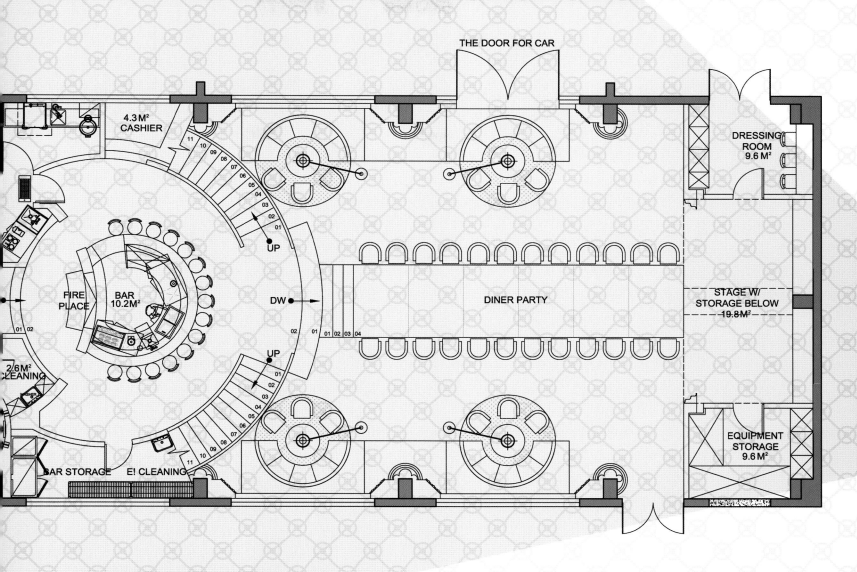

For the constraints of the former internal structure of the old factory with four storeys, the designers decide to only keep the sense of history of the facades but break all the original structures indoor to reconstruct a three-storey building full of line sense. In order to meet the needs of the client to be open, in the process of rebuilding the structure, the designers pay particular attention to building the atrium into a spatial, open area. No matter in which floor, you can communicate with the colleagues in each floor conveniently. The use of the long white piano paint desk not only meets the need of friendly working atmosphere but also provides convenience for employees to communicate.

The design theme of the office is "working in the park". Firstly, the designers choose pure white in the arrangement of the colors and transform the old gloomy factory building into a transparent, bright, concise and pure office area. Secondly, when reconstructing the internal space structure, with the large branches as the inspiration they create a vivid sense of line, which reflects the theme of "working in the park" better. In order to make the whole office space a park which can let people be immersed in, the designers plant the outdoor park concept elements as much as possible into the office, which makes the traditional, rigorous office become fun. On the ground of the office roads lay staggered and the green lawns among them give people a feeling just like working on the lawn; the design of the conference room is based on the tennis court: grid glass wall, plastic ground of the tennis court, table inlaid with tennis; the swing with the color of childhood was placed aside freely, which lets us back to childhood after being tired of the work; on both sides of table some space is set aside for the potted flower deliberately; the wall surface sets upward square drawer to remind people of drawers of flower pots where the plants inside continuously extend upwardly, reflecting the vitality of the company and the function of storage. Here it seems that the work also becomes fun and easy.

While reflecting the fashionable and comfortable office environment atmosphere, the designers do not forget the original function and usage of the office, in which they apply high-tech to reflect the creativity to bring you surprise. When you see a lift, you do not realize that it actually is an autogate to the washroom with the same creativity while you were wondering to go upstairs. The designers use different light colors to distinguish between male and female with pink representing the girls and green representing the boys. Even the smallest detail of the bathroom walls is also based on the graffiti style of the famous French artist and the image in the childhood game as the mosaic pattern. They spare no effort to provide people with the feeling of happiness in any place of the office. In order to make full use of the space, they set an X type shelf at each half platform of the stair in order to benefit the staff upstairs and downstairs to be ready to read at any time, which reflects the cultural atmosphere of the company better.

With the respect for the historical building, the designers keep the complete facade of the building and change the interior structure of the space subtly, which not only carries forward the old industrial temperament of the factory but also complies with the role setting and needs of the client after changing it into a fashionable office area through creative layout and design.

Dunmai 办公室坐落于南外滩的幸福码头时尚创意园区，其前身为上海幸福摩托车厂。园区在车水马龙的南浦大桥地块略显低调。华灯初上的时候，柠黄色的灯光映照在古朴的裙楼间，绵绵绿竹映出婆娑的灯影，勾勒出一处静谧、悠闲、开阔的创意空间。南外滩是上海的标志，也是上海人从20世纪延续至今不灭的情结。老码头、外马路、十六铺……单是看到这些地名，土生土长的上海人脑海中就会缓缓呈现出黄浦江的涛声、氤氲的晨雾，响起巨轮和渡轮汽笛的鸣声。如今的南外滩，不仅延续了老外滩的古雅奢华和董家渡的过往繁荣，也同样代表着外滩和世博园区的科技一体化工程。

这个项目的客户是一家活动策划创意公司，在充分考虑了客户公司的特性和要求后，两位设计师 Thomas Dariel 和 Benoit Arfeuillere 决定将其打造成既符合园区概念的创意办公室，又能带给员工愉悦、轻松氛围的工作空间，并利用高科技的产品来体现办公室的实用性和功能性。

由于之前这座4层的老厂房内部结构的限制，设计师决定只保持有历史感的外立面，并打破室内一切原有结构，重塑一个具有线条感的3层建筑结构。为体现客户开放性的要求，在重塑结构时，设计师特意将中庭打造成一个挑空开放的区域，不论你位于哪一层，都能与其他楼层的同事便捷沟通。白色钢琴漆办公长桌既增添了融洽的工作氛围，又便于员工交流。

本案的设计主题是"在公园里工作"。首先，纯白配色将这个原本阴暗陈旧的老厂房改造成一个通透明亮、简洁纯净的办公区域。其次，在重塑内部空间结构时，以大树枝干为灵感，营造出生动的线条感，更好地体现了"在公园里工作"的主题。为了使员工办公有如置身公园的感觉，设计师尽可能地移入户外公园的概念元素，使传统、严谨的办公室变得妙趣横生。办公室的地面上覆以交错的马路，铺设其中的绿色草坪令人感觉好似在草地上办公；会议室的设计以网球场为雏形，网格的玻璃隔墙、网球场塑胶地面和嵌有网球的桌子放置其中；带有童年色彩的秋千随意地摆放在一旁，工作劳累之时还能回归童年；桌子两边特意留出放置盆栽的花道，使整个办公室春意盎然；墙面上设置向上排列的方格抽屉，让人联想到一个个花盆，种植在里面的植物不断向上延伸，在体现出公司的蓬勃活力之时，又具有一定的储物功能。在这里，工作也似乎变得有趣轻松起来。

在营造时尚轻松的办公环境氛围时，设计师并没有忘记为其打造功能性和实用性。运用高科技来体现创意，让你收获惊喜。当你远远看到一部电梯，正想上楼的瞬间，却意外地发觉这实际是一扇具有创意的卫生间自动门。设计师运用不同的灯光颜色来区别性别，粉色代表女生，绿色代表男生。就连最细节的卫生间墙面，也借鉴了法国著名艺术家的涂鸦风格，采用儿时游戏机中的形象作为马赛克的图案，在任何一处都不遗余力地想要呈现出愉悦的感觉。为了使空间得以充分利用，在每层楼梯的半平台处设有一个"X"形书架，以方便楼上楼下的工作人员随时取阅，更体现了该公司的文化氛围。

怀着对历史建筑的崇敬之情，设计师完好地保留了建筑的外立面，并巧妙地改变内部空间结构，不仅继承了老厂房的工业气质，还通过创意布局和设计使其变成时髦的办公场所，同时符合本案客户的角色定位和需求。

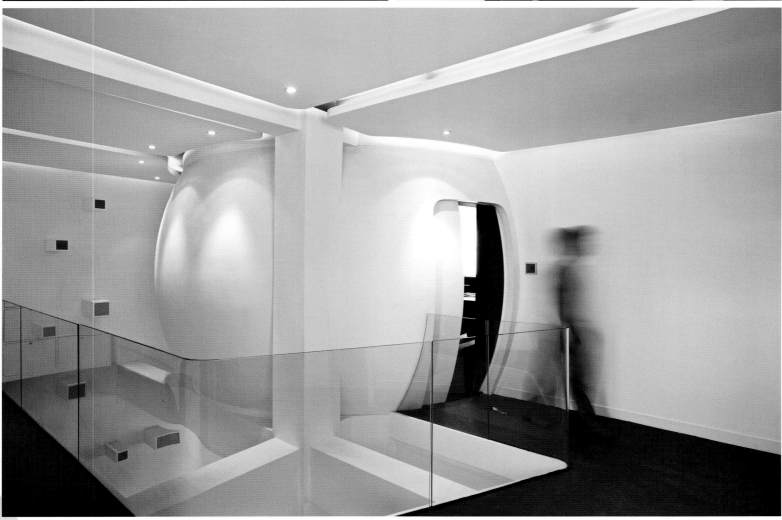

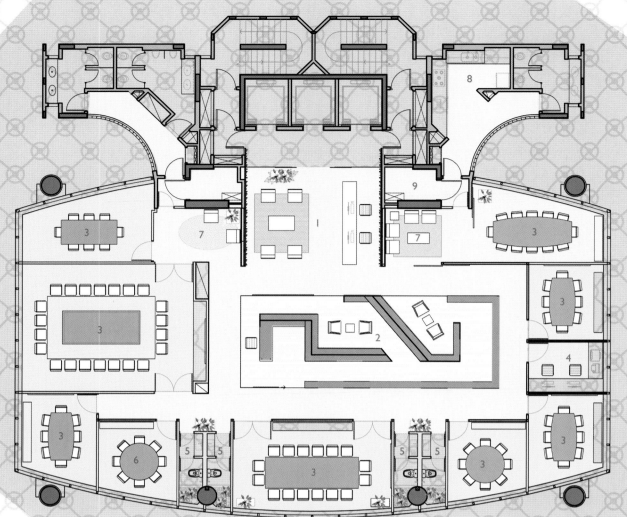

1 - RECEPTION
2 - SUSPENDED LIBRARY
3 - REUNION ROOM
4 - ARCHIVE
5 - BATHROOM
6 - WORKSPACE/MANEGEMENT
7 - WAITING ROOM
8 - KITCHEN
9 - AIR CONDITIONING ROOM

BPGM LAW OFFICE

BPGM 律师事务所

FGMF Arquitetos

项目团队：Ana Beatriz Lima, Bruno Araújo, Marilia Caetano, Marina Almeida
主设计师：Fernando Forte, Lourenço Gimenes, Rodrigo Marcondes Ferraz
合 作 方：Renata Davi
建筑面积：570 平方米
项目地点：巴西圣保罗市

Project Team: Ana Beatriz Lima, Bruno Araújo, Marilia Caetano, Marina Almeida
Principal Designers: Fernando Forte, Lourenço Gimenes, Rodrigo Marcondes Ferraz
Coordinator: Renata Davi
Area: 570 m²
Location: São Paulo, Brazil

Called for a bid of an architecture project for two floors of BPGM Law Office, we came across with a clear division: One floor—larger, operational, and another—slightly smaller, for meeting rooms and administrative area. Once defined the clear division between the functional and social floors, the need to represent the view of BPGM Office to the clients and lawyers fell back on the first one. In the view of the situation, it seemed an interesting option to organize the office in a radial way: every meeting room would be, thus, in the perimeter of the set, with views of the outside and natural illumination, and, in the center of the set—the visitor's focal point—there would be a large library which organizes the flows to the different rooms. The natural illumination of the library and passageways is guaranteed by the continuous glass flags in every divider of the peripheral rooms.

However, the library has earned a great symbolic importance by becoming the organizational element of the plant and flows of the set. It also seemed appropriate to us that it should be the first visible element to the visitor, as soon as one leaves the elevator. Therefore, there was a need that the library was not supposed to be only an element to accumulate and organize books, but also an element to represent the ideology of the office. Its responsibility would become two-fold: a well-organized library, but also an interesting and different element—sober to the point of not compromising the need of seriousness and reliability a law office must show, but innovative and contemporary such as the essence of BPGM Office. There was, then, a sort of conflict between tradition and innovation.

In order to respond to these yearnings we launched ourselves, an idea arose: the library would become a sort of small labyrinth, with unusual angles, and open and closed passages. Besides, some places would have windows where the ones who were in the passageways would be able to look inside, and other places would be completely closed.

We also designed the library so that it would never touch the floor. The whole library became a floating, suspended element that would never touch the flooring, hovering in almost a mysterious way at 40 centimeters of the flagstone. The lawyer's job is, in essence, cerebral. It shapes up in agreements and discussions, but it is a job of services, based on people and knowledge. The library in the center of the set is nothing else than the representation of collective knowledge available to the client of BPGM Office. This abstract knowledge is shaped up in an ethereal, floating way, by not touching the floor or the walls through the element that is the library—a small floating labyrinth which contains the office knowledge and, although contemporary, expresses all its tradition.

F GMF 建筑事务所负责为 BPGM 律师事务所设计两层楼的办公室，建筑师设想：1 层较宽阔，可用于办理日常业务；另一层较小，用作会议室和行政管理区。建筑师清晰地把这两层分为功能区和社会区。一旦明确定义了两个区域的分工后，设计师就开始向客户和员工展示 BPGM 律师事务所的形象。办公室以放射形进行布局：每一个会议室都位于放射形的边缘，既能够看到室外景色又能获得充足的日照。放射形的中心则是访客区，其中有一个大型的图书室，从这里可以通往任意一个办公室。虽然图书室和过道位于中间地带，但是自然光线也能够通过边缘房间的玻璃隔断照射进来。

在整个室内布局当中，图书室具有很重要的象征意义，因为它是访客从电梯出来首先看到的设施。因此，图书室不仅仅用来摆放书籍，还代表了整个律师事务所的思想体系。它具有双重含义：既是一个安排有序的图书室，又是一个独特的元素。它独树一帜，不拘泥于传统律师事务所肃穆端庄的氛围，以新颖的现代姿态呈现出BPGM律师事务所的特质，碰撞出传统与创新的火花。

从这个目的出发，设计师萌生出一个想法，即将图书室设计成一个小型迷宫，设置不同角度、敞开或封闭的通道。另外，一部分走道的墙壁上设有窗户，另一部分则完全封闭，人们可以通过窗户看到内部的事物。

图书室内大部分办公构件都没有触及地面，整个图书室都采用神秘的悬浮式设计，与石地板有40厘米左右的距离。律师的工作本质是理性主义，需时常进行协议与论述。但不应忽略的一点是律师的宗旨是为人服务，应以人与知识为本。办公室中心的图书室布局就是将公共知识分享给BPGM客户的象征。漂浮的图书迷宫，代表了无形的知识缥缈地悬浮于地板墙壁之间，新颖地表达出事务所的传统精神。

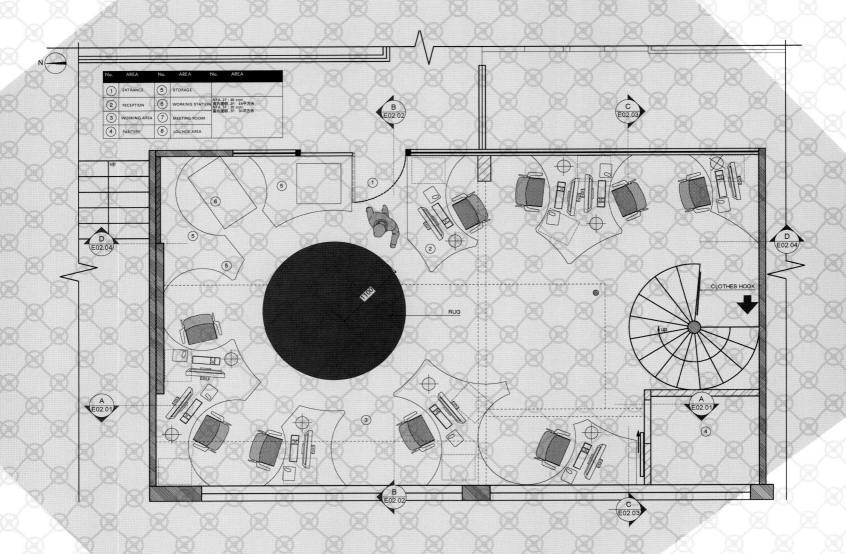

AND-SUPERPRESS-SUPERBLA

AND–SuperPress–SuperBla 办公室

Naco Architectures

设 计 师：Marcelo, Adam Fang
建筑面积：80 平方米
项目地点：中国上海市
摄　　影：徐文磊

Architects: Marcelo, Adam Fang
Area: 80 m²
Location: Shanghai, China
Photography: Wenlei Xu

A new trio of agencies—AND (strategic communication), SuperPress (press and PR), and SuperBla (digital marketing)—in Shanghai were looking for a common office where their avant-garde work methods and ideas could flourish. Designed by Naco Architectures, this 80 m² helps them to do just that. The pure white space makes a bold statement, serving as a blank canvas and the ideal environment for creation. Yet within this pristine white are surprising punches of electric colors. "Colors used in the space are all very energetic, very bright and dense. They will evolve month after month, because the office is not static. Ideas change, colors will change too. They represent splashes of energy, and creativity," explains Naco Shanghai's director, Margaux Lhermitte. The team works in a circle, the arrangement providing each staff with their own space while keeping them connected to the rest of their colleagues.

Nowadays, teams are used to changing projects very fast. In this configuration, changing the teams is very easy, as everybody works on laptops. "We didn't want to create any enclosed or private space in this office, because the team is quite small at present, and brainstorming should be their daily work habit. They should work in teams, and constantly exchange ideas between themselves, and with clients," says Margaux. Margaux adds that the strongest part of the project is—most unexpectedly—the way it looks from the outside, where even as a pure white office it still comes off as warm and welcoming. "It is a small office, but still so many people stop by in the daytime and want to know more about it. This is the best achievement as a designer to make people ask questions," says Margaux.

AND（战略传播）SuperPress（新闻与公关）和 SuperBla（数字营销）这三个品牌组合创立了一个全新的机构，并在上海找到一处可以激发创新工作方法与思路的办公场所。新办公室由 Naco Architectures 设计，在 80 平方米的空间内满足了客户的需求。纯白色的空间如同一张空白画布，为创意的迸发提供了理想的环境。然而在这片纯白的场景中也同样存在着刺激性的颜色。Naco 位于上海分部的设计师 Margaux Lhermitte 认为："这个空间运用的色彩充满活力、明亮而且紧凑，它们将月复一月地更换，因为办公场所并非一成不变。观点在变化，颜色也随之而变，它们代表着飞扬的活力和创造性。" 团队合作将在一个环形区域展开，这种设计不仅让每个员工都拥有自己的空间，同时也使同事之间保持密切沟通。

如今，合作团队总是在高速地变换着项目方案。采用这种布局，小组的沟通形式变得更加简易，因为每个人都用笔记本电脑工作。Margaux 表示："我们并未计划在办公室创造任何密闭或私人空间，因为团队目前规模尚小，自由讨论应当成为日常习惯。员工应该以小组形式工作，与客户和小组成员不断交换意见。"Margaux 补充强调整个项目出人意料的特点，即从外部看来纯白色的办公空间，竟体现出热情好客的风度。"阳光下的办公空间，规模不大，却吸引了众多行人驻足观看、问询了解。让人们产生着恋是设计师的最佳成就。" Margaux 说。

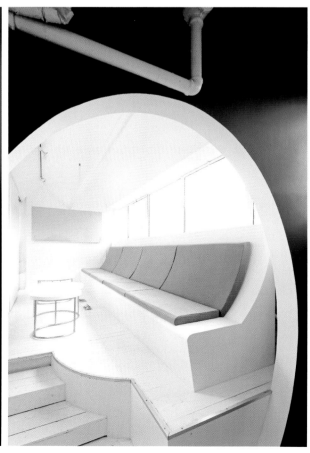

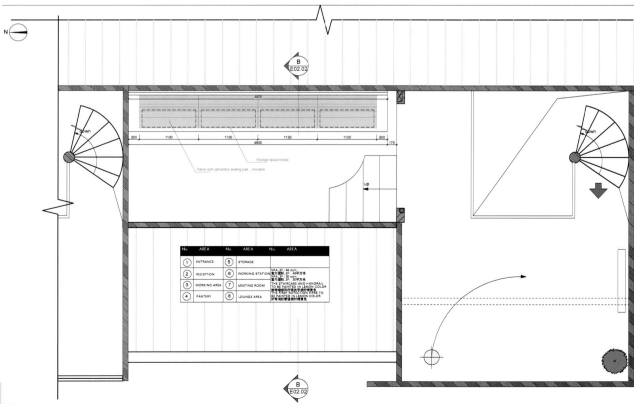

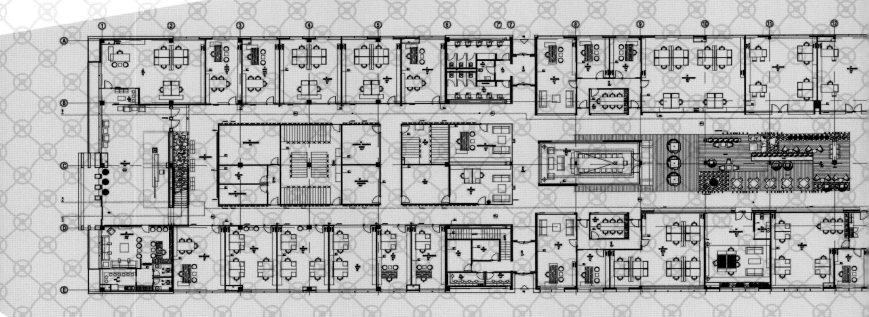

KOZA HOLDING HEADQUARTERS
Koza 控股总部

Craft312 Studio

设 计 师：Onur Karlidag, Deniz Karlidag
建筑面积：6 000 平方米
项目地点：土耳其安卡拉市
摄　　影：Nur Acar Tangun

Architects: Onur KARLIDAG, Deniz KARLIDAG
Area: 6,000 m²
Location: Ankara, Turkey
Photography: Nur Acar Tangun

Koza Holding Headquarters, which has 6,000 m² used area and with the number of 300 working person, is designed as a three compartments; entrance and information area, inner garden and the offices, design studios and refectory. 200 meters long building as a horizontal skyscraper is a factory building by peeling the walls and becoming only its columns and beams for designing new functions.

Prestige and identity to reflect concerns from the entrance is kept in the forefront of design concept, granite and wooden panel space with a modern interpretation and is reflected in a large reception desk. This approach is dominated by linear editing, design elements and other details need compartments to emphasize ecology. Intensive use creates a comfortable circulation and orientation guide, the linear lighting elements across the hallway entrance also bring a sense of continuity.

Above the administration floor doubled the welcome section connecting the stairs, paneled wall hidden behind the reception desk and security has been a kind of separation. In administration floor, 9 m-long executive assistant desk is connecting to corridors which are dominated by shades of brown in dramatic light way to be created in the transition areas.

Video wall and the second standby pass the entrance corridor, the L-shaped corridor wooden panels with hidden lighting, continuation of the middle parts of the garden and the café. The middle inner garden of employees working in offices in green and seeing and feeling nature is based on the theme of a "Green Office" standing out as a scenario. Inner Garden was opened on the glass of the building through cracks and gaps in the light of day directly to the ceiling grid system, with receiving and distributing all the space. Inside the cafe in the garden, gives employees a place that can break, relax, socialize and feel the light of day. In order to support this fiction, the middle area and the indoor plants used in the landscape design supported by a water curtain are designed to reduce the power of all the day stress and noise from corridors.

Offices are opened to inner garden and use day light in every hour in every day efficiently. In the internal garden offices, work stations, convenience functional sense meeting facilities, libraries, and storage units made of acrylic and wood, and specifically to the needs of today's modern offices were designed by Craft312 Studio. In offices located in sensitive ecological building materials, longevity and understanding of interior design solutions presented in the foreground. The correct use of light and working conditions to meet the expectations of comfort and functionality, from offices in the light of day, supplemented with task lighting over lighting, and hidden lights. Finally, whole 6,000 m2 office spaces are designed for creating a new ecological working environment era and interior design project with a unique way of using day lights, materials and plants.

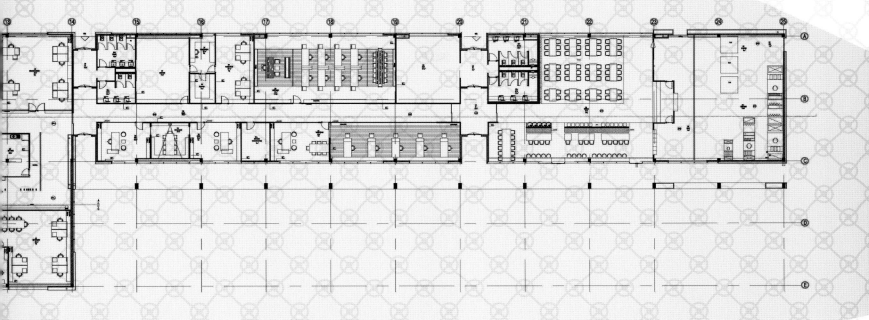

Koza 控股总部拥有 6 000 平方米使用面积，可容纳 300 名员工。它被设计成三个部分：入口和信息区域、内院和办公室、设计工作室和餐厅。其建筑为 200 米长的横向摩天大厦，仅利用剥落的厂房墙壁、立柱和横梁来设计全新的功能。

入口处秉承最前沿的设计理念，设置宽阔的前台，用花岗岩及木板空间诠释现代风格，处处体现着其社会威望和良好的企业形象。这种方法主要采用线性形式编辑设计元素，空间分隔所需的其他细节注重生态性。集约利用创造了舒适的工作环境和空间归属感，整个走廊入口的线性照明元素也带来了连续感。

行政部上层连接楼梯的两个接待点，隐蔽在接待台后方的墙板起着分隔和保持一定私密安全的作用。在行政楼层，9 米长的执行助理办公桌连接着散发棕色色调的走廊，同时在过渡区采用戏剧性光效手法。

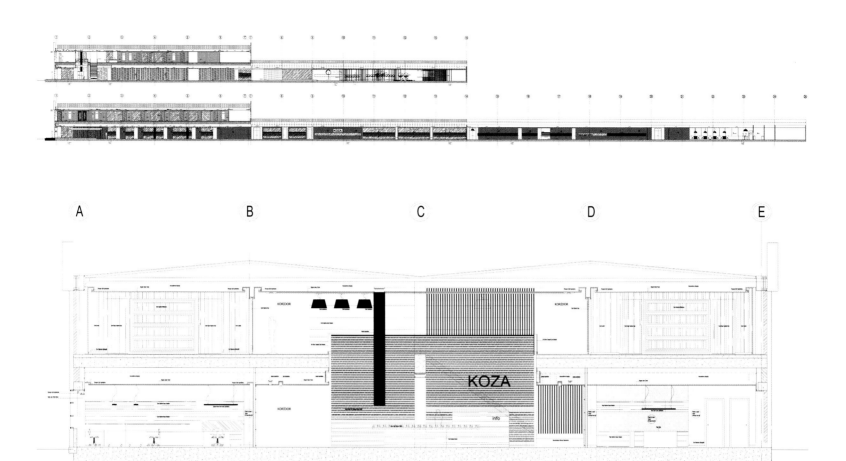

电视墙和第二套备用设备穿过入口通道，内置式照明隐藏在 L 形走廊的木板里，走廊成为了花园中心区和咖啡厅的延续。员工办公室的中部内院种有绿植，人们可以身处其中体会大自然的美妙，这突出了以"绿色办公室"为主题的办公氛围。内院可以通过建筑物的玻璃呈现出来，白天光线可以直接通过裂缝和开口直射天花板的网格系统，这一系统将光线吸收并折射到整片区域。花园咖啡厅为员工提供了休息、放松、社交和享受阳光的场所。为支持这一功能，中部区域和室内的绿植采用水幕形式，帮助员工释放一整天的压力，同时减少来自走廊的噪音。

办公室通向内花园，可以每时每刻地有效利用日光。Craft312 设计工作室在内部花园办公室中，精心设计工作区、便利的功能会议设施、图书馆、木制储物柜，甚至是其他一些满足当今现代化办公需求的区域。办公室由感光的生态建材建造，并前瞻性地提出了长久耐用的室内解决方案。设置办公室日用灯光系统、辅助照明系统和灯光隐藏系统，合理应用灯光，创造满足功能性和舒适性的工作条件。最后，通过对整个 6 000 平方米办公空间的设计，开创了一个重视生态化工作环境的新时代，完成了一个采用自然光源、材料和植物的独特室内设计项目。

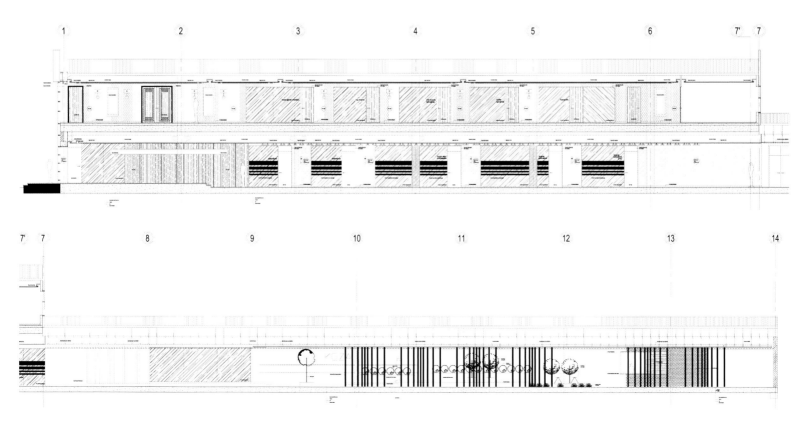

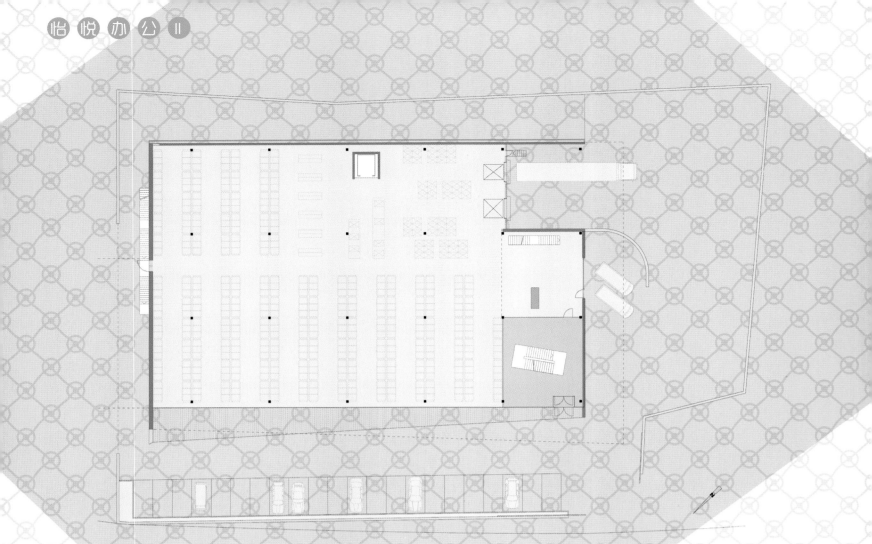

ROTHOBLAAS LIMITED COMPANY

RothoBlaas 公司总部办公楼

Monovolume Architecture+Design

项目总监：Pedó Pobitzer
项目团队：Christian Gold, Barbara Waldboth, Amgelika Mair
客　　户：Rotho Blaas Limited Company
建筑面积：3 700 平方米
项目地点：意大利阿尔托·阿迪杰地区
摄　　影：Oskar Da Riz

Project Manager: Pedó Pobitzer
Project Team: Christian Gold, Barbara Waldboth, Amgelika Mair
Client: Rotho Blaas Limited Company
Area: 3,700 m²
Location: Kurtatsch, Italy
Photography: Oskar Da Riz

The "RothoBlaas" office is a large-scale commercial operation specializing in assembling systems and power tools for the woodworking industry. Warehouse and commissioning are situated on the ground floor whereas administration, a meeting room and a showroom can be found on the upper floor. The aim of the project was to create a compact building with a high level of recognition, the building as corporate identity of the enterprise is contemporary and representative among the companies. This has led to a functional, compact structural shell, provided with a glass envelope. The main building material employed is wood in order to show the company's own products.

"RothoBlaas"公司是一家主要为木制品行业提供组装系统和电动工具的大型商业机构。其仓库和调试室位于底层，而行政处、会议室和陈列室则在较高的楼层。项目目的在于建造一座众所周知的紧密型建筑，其形象具有其他公司无可比拟的时代性与代表性。因此设计师们在建筑表层之上放置了一张功能齐全、结构紧凑的玻璃罩结构。建筑内部主要采用木质原材料，用以展示他们自己的产品。

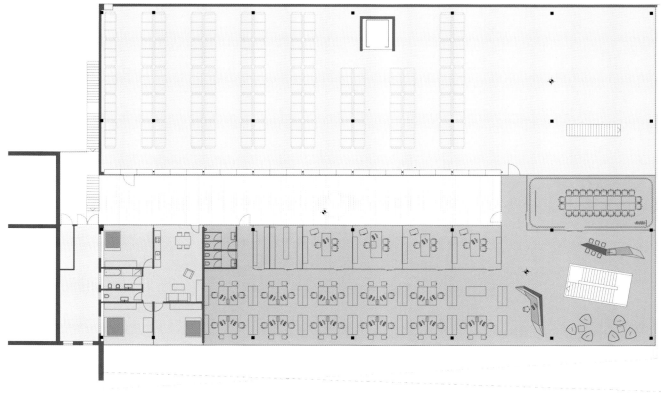

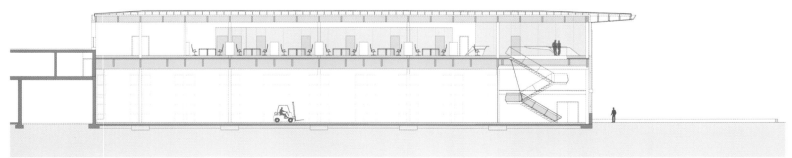

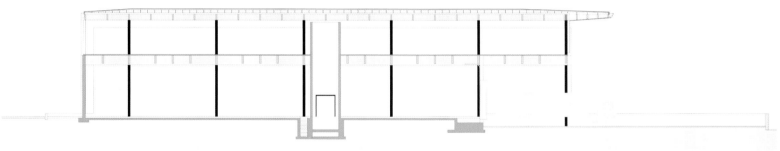

RED BULL AMSTERDAMR HEADQUARTERS

红牛阿姆斯特丹总部大厦

Sid Lee Architecture

视觉平面设计：Sid Lee
客　　户：Red Bull Netherlands
建筑面积：875 平方米
项目地点：荷兰阿姆斯特丹市
摄　　影：Ewout Huibers

Visual Identity and Graphics: Sid Lee
Clients: Red Bull Netherlands
Area: 875 m²
Location: Amsterdam, the Netherlands
Photography: Ewout Huibers

Montreal, September 16, 2010 – chosen over two other firms, Sid Lee Architecture and Sid Lee's Amsterdam atelier were mandated to create the new Red Bull Amsterdam headquarters. The company agreed to settle in the North side of Amsterdam's port area, in a site evocative of both an artistic street culture and the intensity of extreme sports. The project landed in an old heritage shipbuilding factory, facing a timeless crane and an old disused Russian submarine.

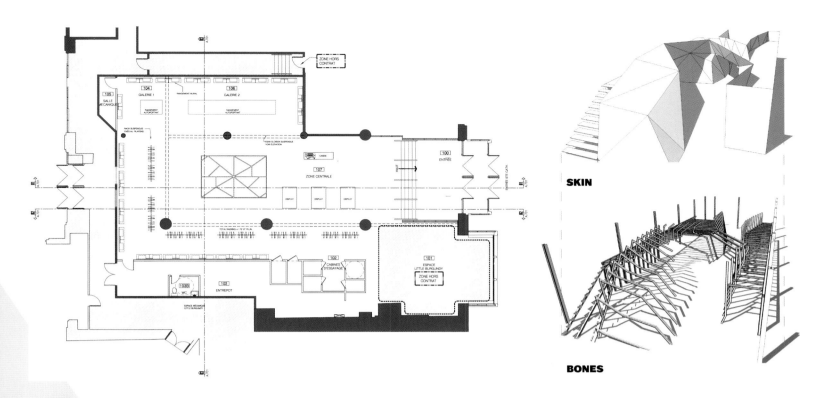

SKIN

BONES

Architecture to suit a philosophy
"To design the inner space, we aimed at retrieving Red Bull's philosophy, dividing spaces according to their use and spirit, to suggest the idea of the two opposed and complementary hemispheres of the human mind, reason versus intuition, arts versus the industry, dark versus light, the rise of the angel versus the mention of the beast", says Jean Pelland, lead design architect and senior partner at Sid Lee Architecture.
Inside the shipbuilding factory, with its three adjacent bays, the architects focused on expressing the dichotomy of space, shifting from public spaces to private ones, from black to white and from white to black. Our goal in this endeavor was to combine the almost brutal simplicity of an industrial built with Red Bull's mystical invitation to perform. The interior architecture with its multiple layers of meaning conveys this dual personality, reminding the user of mountain cliffs one moment and skate board ramps the next. These triangle-shaped piles, as if ripped off the body of a ship, build up semi-open spaces that can be viewed from below, as niches, or from above, as bridges and mezzanines spanning across space. In the architecture we offer, nothing is clearly set; all is a matter of perception.

在2010年9月16日蒙特利尔市，Sid Lee 建筑事务所与 Sid Lee 阿姆斯特丹画室在与另外两家候选公司的设计竞标中脱颖而出，被红牛阿姆斯特丹选中建造他们新的总部大楼。红牛同意将新总部设立在阿姆斯特丹港口区的北部，这是一个街头艺术文化、极限运动文化盛行的区域。项目建在一间古老造船厂的旧址上，工厂里面至今仍保留着一个老旧的起重机和废弃的古老俄罗斯潜艇。
建筑的哲学
Sid Lee 建筑事务所的首席设计师兼资深合伙人珍·佩兰德说道："在室内空间的设计上，我们以唤起红牛哲学为目标，按照功能与精神的方式进行划分，以人类左右大脑的对立性与互补性为题，提炼出理性与直觉、艺术与工业、黑暗与光明、天使与野兽的对比性。"
造船工厂内部如同三个相邻的海湾，建筑师集中表达了二分法的观点，即从公共空间向私密空间的转换、从黑到白再到黑的转变。在这种尝试里，建筑师的目标是将工业建筑极致的简洁性与红牛神秘的诱惑力结合在一起。内部建筑上采用多层次手法表达这一双重性，提醒使用者此一时彼一时。这些三角形的堆积宛如没有船身的船只，逐步建构出一个半开放性的空间。由下往上的视觉感受如同壁龛，由上往下则像是桥梁或跨越式夹层空间。建筑师们所提供的建筑方案不是对事物的清晰设定，只是一种对事物的感知。

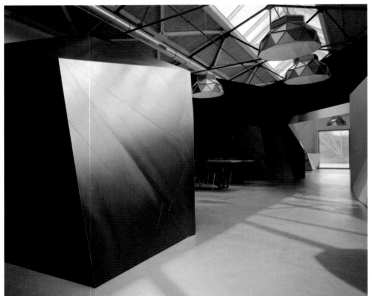

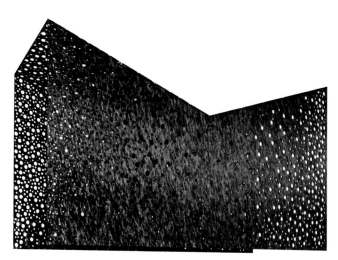
METAL SHELL

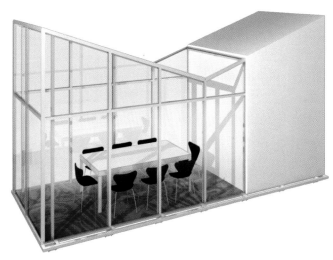
GLASS ENCLOSURE

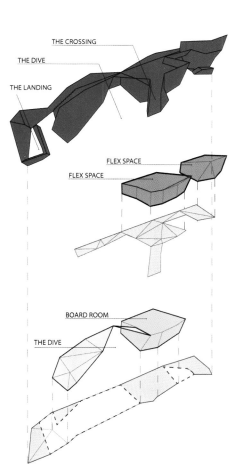

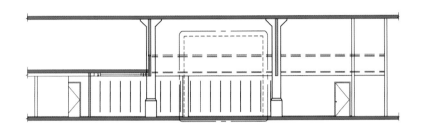

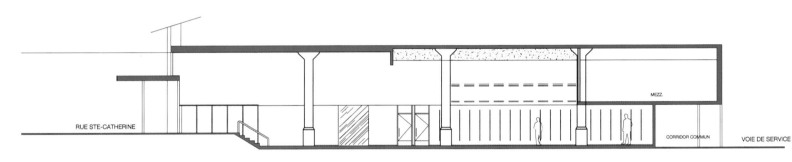

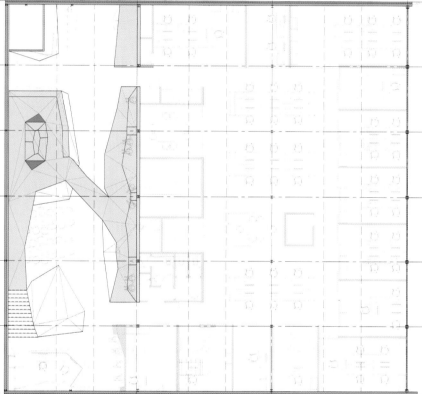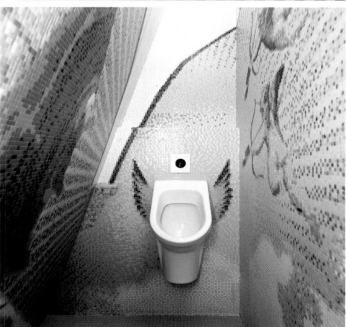

图书在版编目（CIP）数据

怡悦·办公Ⅱ / 深圳市博远空间文化发展有限公司编.
— 天津：天津大学出版社，2013.3
ISBN 978-7-5618-4557-8

Ⅰ.①怡… Ⅱ.①深… Ⅲ.①办公室－室内装饰设计－作品集－世界－现代 Ⅳ.①TU238

中国版本图书馆CIP数据核字（2012）第284601号

怡悦·办公 Ⅱ　　　　　博远空间文化发展有限公司 编

版式设计：赵童
责任编辑：郝永丽
出版发行：天津大学出版社
出 版 人：杨欢
电　　话：发行部：022-27403647
网　　址：publish.tju.edu.cn
地　　址：天津市卫津路92号天津大学内（邮编：300072）
经　　销：全国新华书店
印　　刷：深圳市彩美印刷有限公司
开　　本：235 mm × 320 mm
印　　张：20
字　　数：298千字
版　　次：2013年3月第1版
印　　次：2013年3月第1次印刷
书　　号：ISBN 978-7-5618-4557-8
定　　价：320.00元（USD 59.90）

（本图书凡属印刷、装帧错误，可向发行公司调换）

AFTERWORD

后 记

"一切的变化都可以归结为时空的变化,设计师要面对的是改变了时空的技术和被时空改变的人",这句话可以生动地概括办公空间的发展过程和设计趋势。现代新型办公空间与原有的传统办公空间相比发生了许多重要变化,由于信息技术的发展、公司组织结构的水平化变动和人文精神的"苏醒",各种因素综合要求新型办公空间能够及时应对环境的变化,这种环境可以是经济的、管理的、工作的或者情感的等诸多方面的,这就要求办公空间和企业结构、办公程序组成一个紧密的系统整体,不再是以往的彼此脱节的松散结构。同时,在系统化的前提下呈现多元化的形态特征,满足不同状态的工作需要。优秀的办公空间设计可以归纳为交流性——信息交流的场所,社会性——办公空间结构的社会化表达,流动性——办公空间的动态适应性发展三个主要方面的特征。

办公空间设计是一个系统化的设计过程,我们视觉所及的设计内容只占整个设计过程很小的一部分,它包括基础设施、各个功能分区和整体氛围等诸多方面,当然也包括空间具体细化尺寸设计和光、声、热环境设计等内容,但是最根本的是"以人为本"的建筑设计精神,是带有深层次人文关怀的人性化理念。

本书秉承《怡悦办公1》的严谨态度和精益求精的精神,经过无数个日夜的锤炼,终于跟读者见面。精诚所至,金石为开。希望这本《怡悦办公2》能够延续辉煌并有所突破,给读者带来更多的惊喜。当然,世上没有完美之事,若有不当之处,敬请读者批评和指正。在以后的日子里,我们将以加倍的努力,为您奉上更多精彩图书,敬请期待!

"Every change can be attributed to time and space changes, and the designers have to face the technology which has changed the time and space and the people who have been changed by the time and space", this sentence can summarize the development process and design trend of the office space vividly. Compared with the original and traditional office space, the modern office space has taken a lot of significant changes. Due to the development of information technology, the level change of company's organization structure and the "awakening" of humanistic spirit, the combination of various factors require the new office space to be adjusted to the environment, which can be economy, management, work or emotion, etc.. Then it requires the office space, company structure and working procedures to form a close system but not be a disjointed loose structure in the past. At the same time, in the premise of system it appears to be diversified characteristics to meet different needs of working conditions. The excellent office space design can be summed up as the following three aspects: communication—place for information exchange, sociality—social expression of the office space structure, liquidity—dynamic adaptability development of the office space development.

Office space design is a systematic design process, the design content in our vision only accounts for a very small part of the entire design process, which includes not only infrastructure, each functional area, the whole atmosphere and so on but also the specific and detailed design size of space, acoustic, sound, thermal environment and so on. But among them the most important thing is the "people-oriented" spirit of the architectural design and the humanization concept with deep humanistic concern.

The book *Joy Office 1* upholds strict attitude and the spirit of striving for perfection continuously and comes to meet the readers after days and nights of efforts. No difficulty is insurmountable if one sets one's mind on it. We hope that this *Joy Office 2* will carry forward the resplendence and have its own breakthrough to bring more surprises for the readers. Of course, there is no perfect thing in the world. If there is something inappropriate, you are welcomed to criticize and point out the mistakes. In the future we will redouble our efforts to bring you with more wonderful books. So please stay tuned!